Bayesian Statistical Methods

CHAPMAN & HALL/CRC
Texts in Statistical Science Series

Joseph K. Blitzstein, *Harvard University, USA*
Julian J. Faraway, *University of Bath, UK*
Martin Tanner, *Northwestern University, USA*
Jim Zidek, *University of British Columbia, Canada*

Recently Published Titles

For more information about this series, please visit: https://www.crcpress.com/go/textsseries

Bayesian Statistical Methods

Brian J. Reich
Sujit K. Ghosh

CRC Press
Taylor & Francis Group
Boca Raton London New York

CRC Press is an imprint of the
Taylor & Francis Group, an **informa** business

A CHAPMAN & HALL BOOK

Chapman & Hall/CRC Press
Taylor & Francis Group
6000 Broken Sound Parkway NW, Suite 300
Boca Raton, FL 33487-2742

First issued in paperback 2021

© 2019 by Taylor & Francis Group, LLC
CRC Press is an imprint of Taylor & Francis Group, an Informa business

No claim to original U.S. Government works

ISBN 13: 978-1-03-209318-5 (pbk)
ISBN 13: 978-0-815-37864-8 (hbk)

Publisher's Note
The publisher has gone to great lengths to ensure the quality of this reprint but points out that some imperfections in the original copies may be apparent.

Visit the Taylor & Francis Web site at
http://www.taylorandfrancis.com

and the CRC Press Web site at
http://www.crcpress.com

To Michelle, Sophie, Swagata, and Sabita

Contents

Preface

Bayesian methods are standard in various fields of science including biology, engineering, finance and genetics, and thus they are an essential addition to an analyst's toolkit. In this book, we cover the material we deem indispensable for a practicing Bayesian data analyst. The book covers the most common statistical methods including the t-test, multiple linear regression, mixed models and generalized linear models from a Bayesian perspective and includes many examples and code to implement the analyses. To illustrate the flexibility of the Bayesian approach, the later chapters explore advanced topics such as nonparametric regression, missing data and hierarchical models. In addition to these important practical matters, we provide sufficient depth so that the reader can defend his/her analysis and argue the relative merits of Bayesian and classical methods.

The book is intended to be used as a one-semester course for advanced undergraduate and graduate students. At North Carolina State University (NCSU) this book is used for a course comprised of undergraduate statistics majors, non-statistics graduate students from all over campus (e.g., engineering, ecology, psychology, etc.) and students from the Masters of Science in Statistics Program. Statistics PhD students take a separate course that covers much of the same material but at a more theoretical and technical level. We hope this book and associated computing examples also serve as a useful resource to practicing data analysts. Throughout the book we have included case studies from several fields to illustrate the flexibility and utility of Bayesian methods in practice.

It is assumed that the reader is familiar with undergraduate-level calculus including limits, integrals and partial derivatives and some basic linear algebra. Derivation of some key results are given in the main text when this helps to communicate ideas, but the vast majority of derivations are relegated to the Appendix for interested readers. Knowledge of introductory statistical concepts through multiple regression would also be useful to contextualize the material, but this background is not assumed and thus not required. Fundamental concepts are covered in detail but with references held to a minimum in favor of clarity; advanced concepts are described concisely at a high level with references provided for further reading.

The book begins with a review of probability in the first section of Chapter 1. A solid understanding of this material is essential to proceed through the book, but this section may be skipped for readers with the appropriate background. The remainder of Chapter 1 through Chapter 5 form the core of the book. Chapter 1 introduces the central concepts of and motivation for Bayesian inference. Chapter 2 provides ways to select the prior distribution which is the genesis of a Bayesian analysis. For all but the most basic methods, advanced computing tools are required, and Chapter 3 covers these methods with the most weight given to Markov chain Monte Carlo which is used for the remainder of the book. Chapter 4 applies these tools to common statistical models including multiple linear regression, generalized linear models and mixed effects models (and more complex regression models in Section 4.5 which may be skipped if needed). After cataloging numerous statistical methods in Chapter 4, Chapter 5 treats the problem of selecting an appropriate model for a given dataset and verifying and validating that the model fits the data well. Chapter 6 introduces hierarchical modeling as a general framework to extend Bayesian methods to complex problems, and illustrates this approach using detailed case studies. The final chapter investigates the theoretical properties of Bayesian methods, which is important to justify their use but can be omitted if the course is intended for non-PhD students.

The choice of software is crucial for any modern textbook or statistics course. We elected to use R as the software platform due to its immense popularity in the statistics community, and access to online tutorials and assistance. Fortunately, there are now many options within R to conduct a Bayesian analysis and we compare several including JAGS, BUGS, STAN and NIMBLE. We selected the package JAGS as the primary package for no particularly strong reason other than we have found it works well for the courses taught at our university. In our assessment, JAGS provides the nice combination of ease of use, speed and flexibility for the size and complexity of problems we consider. A repository of code and datasets used in the book is available at

https://bayessm.org/.

Throughout the book we use R/JAGS, but favor conceptual discussions over computational details and these concepts transcend software. The course webpage also includes latex/beamer slides.

We wish to thank our NCSU colleagues Kirkwood Cloud, Qian Guan, Margaret Johnson, Ryan Martin, Krishna Pacifici and Ana-Maria Staicu for providing valuable feedback. We also thank the students in Bayesian courses taught at NCSU for their probing questions that helped shape the material in the book. Finally, we thank our families and friends for their enduring support, as exemplified by the original watercolor painting by Swagata Sarkar that graces the front cover of the book.

1

Basics of Bayesian inference

CONTENTS

1.1   Probability background

Understanding basic probability theory is essential to any study of statistics. Generally speaking, the field of probability assumes a mathematical model for the process of interest and uses the model to compute the probability of events (e.g., what is the probability of flipping five straight heads using a fair coin?). In contrast, the field of statistics uses data to refine the probability model and test hypotheses related to the underlying process that generated the data (e.g., given we observe five straight heads, can we conclude the coin is biased?). Therefore, probability theory is a key ingredient to a statistical analysis, and in this section we review the most relevant concepts of probability for a Bayesian analysis.

Before developing probability mathematically, we briefly discuss probability from a conceptual perspective. The objective is to compute the probability of an event, $\mathcal{A}$, denoted $\text{Prob}(\mathcal{A})$. For example, we may be interested in the probability that the random variable X (random variables are generally represented with capital letters) takes the specific value x (lower-case letter), denoted $\text{Prob}(X = x)$, or the probability that X will fall in the interval $[a, b]$, denoted $\text{Prob}(X \in [a, b])$. There are two leading interpretations of this statement: objective and subjective. An objective interpretation views $\text{Prob}(\mathcal{A})$ as a purely mathematical statement. A frequentist interpretation is that if we repeated the experiment many times and recorded the sample proportion of the times $\mathcal{A}$ occurred, this proportion would eventually converge to the number $\text{Prob}(\mathcal{A}) \in [0, 1]$ as the number of samples increases. A subjective interpretation is that $\text{Prob}(\mathcal{A})$ represents an individual's degree of belief, which is often quantified in terms of the amount the individual would be willing to wager that $\mathcal{A}$ will occur. As we will see, these two conceptual interpretations of probability parallel the two primary statistical frameworks: frequentist and Bayesian. However, a Bayesian analysis makes use of both of these concepts.

1.1.1 Univariate distributions

The random variable X's support $\mathcal{S}$ is the smallest set so that $X \in \mathcal{S}$ with probability one. For example, if X the number of successes in n trials then $\mathcal{S} = \{0, 1, ..., n\}$. Probability equations differ based on whether $\mathcal{S}$ is a countable set: X is a discrete random variable if $\mathcal{S}$ is countable, and X is continuous otherwise. Discrete random variables can have a finite (rainy days in the year) or an infinite (number lightning strikes in a year) number of possible outcomes as long as the number is countable, e.g., a random count $X \in \mathcal{S} = \{0, 1, 2, ...\}$ has an infinite but countable number of possible outcomes and is thus discrete. An example of a continuous random variable is the amount of rain on a given day which can be any real non-negative number and so $\mathcal{S} = [0, \infty)$.

1.1.1.1 Discrete distributions

For a discrete random variable, the probability mass function (PMF) $f(x)$ assigns a probability to each element of X's support, that is,

$$\text{Prob}(X = x) = f(x). \tag{1.1}$$

A PMF is valid if all probabilities are non-negative, $f(x) \geq 0$, and sum to one, $\sum_{x \in \mathcal{S}} f(x) = 1$. The PMF can also be used to compute probabilities of more complex events by summing over the PMF. For example, the probability that X is either x_1 or x_2, i.e., $X \in \{x_1, x_2\}$, is

$$\text{Prob}(X = x_1 \text{ or } X = x_2) = f(x_1) + f(x_2). \tag{1.2}$$

Generally, the probability of the event that X falls in a set $\mathcal{S}' \subset \mathcal{S}$ is the sum over elements in $\mathcal{S}'$,

$$\text{Prob}(X \in \mathcal{S}') = \sum_{x \in \mathcal{S}'} f(x). \tag{1.3}$$

Using this fact defines the cumulative distribution function (CDF)

$$F(x) = \text{Prob}(X \le x) = \sum_{c \le x} f(c). \tag{1.4}$$

A PMF is a function from the support of X to the probability of events. It is often useful to summarize the function using a few interpretable quantities such as the mean and variance. The expected value or mean value is

$$\text{E}(X) = \sum_{x \in S} x f(x) \tag{1.5}$$

and measures the center of the distribution. The variance measures the spread around the mean via the expected squared deviation from the center of the distribution,

$$\text{Var}(X) = \text{E}\{[X - \text{E}(X)]^2\} = \sum_{x \in S} [x - E(X)]^2 f(x). \tag{1.6}$$

The variance is often converted to the standard deviation $\text{SD}(X) = \sqrt{\text{Var}(X)}$ to express the variability on the same scale as the random variable.

The central concept of statistics is that the PMF and its summaries (such as the mean) describe the population of interest, and a statistical analysis uses a sample from the population to estimate these functions of the population. For example, we might take a sample of size n from the population. Denote the ith sample value as $X_i \sim f$ ("$\sim$" means "is distributed as"), and $X_1, ..., X_n$ as the complete sample. We might then approximate the population mean $\text{E}(X)$ with the sample mean $\bar{X} = \frac{1}{n} \sum_{i=1}^{n} X_i$, the probability of an outcome $f(x)$ with the sample proportion of the n observations that equal x, and the entire PMF $f(x)$ with a sample histogram. However, even for a large sample, $\bar{X}$ will likely not equal $\text{E}(X)$, and if we repeat the sampling procedure again we might get a different $\bar{X}$ while $\text{E}(X)$ does not change. The distribution of a statistic, i.e., a summary of the sample, such as $\bar{X}$ across random samples from the population is called the statistic's sampling distribution.

A statistical analysis to infer about the population from a sample often proceeds under the assumption that the population belongs to a parametric family of distributions. This is called a parametric statistical analysis. In this type of analysis, the entire PMF is assumed to be known up to a few unknown parameters denoted $\boldsymbol{\theta} = (\theta_1, ..., \theta_p)$ (or simply θ if there is only $p = 1$ parameter). We then denote the PMF as $f(x|\boldsymbol{\theta})$. The vertical bar "|" is read as "given," and so $f(x|\boldsymbol{\theta})$ gives the probability that $X = x$ given the parameters $\boldsymbol{\theta}$. For example, a common parametric model for count data with $S = \{0, 1, 2, ...\}$ is the Poisson family. The Poisson PMF with unknown parameter $\theta \ge 0$ is

$$\text{Prob}(X = x|\theta) = f(x|\theta) = \frac{\exp(-\theta)\theta^x}{x!}. \tag{1.7}$$

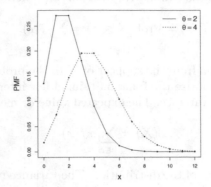

FIGURE 1.1
Poisson probability mass function. Plot of the PMF $f(x|\theta) = \frac{\exp(-\theta)\theta^x}{x!}$
for $\theta = 2$ and $\theta = 4$. The PMF is connected by lines for visualization, but the
probabilities are only defined for $x = \{0, 1, 2, ...\}$.

Clearly $f(x|\theta) > 0$ for all $x \in \mathcal{S}$ and it can be shown that $\sum_{x=0}^{\infty} f(x|\theta) = 1$
so this is a valid PMF. As shown in Figure 1.1, changing the parameter θ
changes the PMF and so the Poisson is not a single distribution but rather a
family of related distributions indexed by θ.

A parametric assumption greatly simplifies the analysis because we only
have to estimate a few parameters and we can compute the probability of any
x in $\mathcal{S}$. Of course, this assumption is only useful if the assumed distribution
provides a reasonable fit to the observed data, and thus a statistician needs
a large catalog of distributions to be able to find an appropriate family for a
given analysis. Appendix A.1 provides a list of parametric distributions, and
we discuss a few discrete distributions below.

Bernoulli: If X is binary, i.e., $\mathcal{S} = \{0, 1\}$, then X follows a Bernoulli(θ)
distribution. A binary random variable is often used to model the result of a
trial where a success is recorded as a one and a failure as zero. The parameter
$\theta \in [0, 1]$ is the success probability $\text{Prob}(X = 1|\theta) = \theta$, and to be a valid PMF
we must have the failure probability $\text{Prob}(X = 0|\theta) = 1 - \theta$. These two cases
can be written concisely as

$$f(x|\theta) = \theta^x (1 - \theta)^{1-x}. \tag{1.8}$$

This gives mean

$$\text{E}(X|\theta) = \sum_{x=0}^{1} x f(x|\theta) = f(1|\theta) = \theta \tag{1.9}$$

and variance $\text{Var}(X|\theta) = \theta(1 - \theta)$.

Binomial: The binomial distribution is a generalization of the Bernoulli to

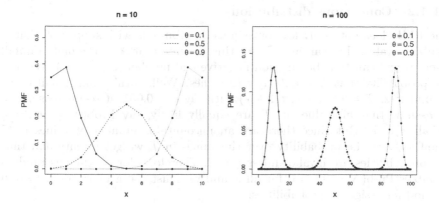

FIGURE 1.2
Binomial probability mass function. Plot of the PMF $f(x|\theta) = \binom{n}{x}\theta^x(1-\theta)^{n-x}$ for combinations of the number of trials n and the success probability θ. The PMF is connected by lines for visualization, but the probabilities are only defined for $x = \{0, 1, ..., n\}$.

the case of $n \geq 1$ independent trials. Specifically, if $X_1, ..., X_n$ are the binary results of the n independent trials each with success probability θ (so that $X_i \sim \text{Bernoulli}(\theta)$ for all $i = 1, ..., n$) and $X = \sum_{i=1}^{n} X_i$ is the total number of successes, then X's support is $\mathcal{S} = \{0, 1, ..., n\}$ and $X \sim \text{Binomial}(n, \theta)$. The PMF is

$$f(x|\theta) = \binom{n}{x}\theta^x(1-\theta)^{n-x}, \tag{1.10}$$

where $\binom{n}{x}$ is the binomial coefficient. This gives mean and variance $\text{E}(X|\theta) = n\theta$ and $\text{Var}(X|\theta) = n\theta(1-\theta)$. It is certainly reasonable that the expected number of successes in n trials is n times the success probability for each trial, and Figure 1.2 shows that the variance is maximized with $\theta = 0.5$ when the outcome of each trial is the least predictable. Appendix A.1 provides other distributions with support $\mathcal{S} = \{0, 1, ..., n\}$ including the discrete uniform and beta-binomial distributions.

Poisson: When the support is the counting numbers $\mathcal{S} = \{0, 1, 2, ...\}$, a common model is the Poisson PMF defined above (and plotted in Figure 1.1). The Poisson PMF can be motivated as the distribution of the number of events that occur in a time interval of length T if the events are independent and equally likely to occur at any time with rate θ/T events per unit of time. The mean and variance are $\text{E}(X|\theta) = \text{Var}(X|\theta) = \theta$. Assuming that the mean and variance are equal is a strong assumption, and Appendix A.1 provides alternatives with support $\mathcal{S} = \{0, 1, 2, ...\}$ including the geometric and negative binomial distributions.

1.1.1.2 Continuous distributions

The PMF does not apply for continuous distributions with support $\mathcal{S}$ that is a subinterval of the real line. To see this, assume that X is the daily rainfall (inches) and can thus be any non-negative real number, $\mathcal{S} = [0, \infty)$. What is the probability that X is exactly $\pi/2$ inches? Well, within some small range around $\pi/2$, $\mathcal{T} = (\pi/2 - \epsilon, \pi/2 + \epsilon)$ with say $\epsilon = 0.001$, it seems reasonable to assume that all values in $\mathcal{T}$ are equally likely, say $\text{Prob}(X = x) = q$ for all $x \in \mathcal{T}$. But since there are an uncountable number of values in $\mathcal{T}$ when we sum the probability over the values in $\mathcal{T}$ we get infinity and thus the probabilities are invalid unless $q = 0$. Therefore, for continuous random variables $\text{Prob}(X = x) = 0$ for all x and we must use a more sophisticated method for assigning probabilities.

Instead of defining the probability of outcomes directly using a PMF, for continuous random variables we define probabilities indirectly through the cumulative distribution function (CDF)

$$F(x) = \text{Prob}(X \le x). \tag{1.11}$$

The CDF can be used to compute probabilities for any interval, e.g., in the rain example $\text{Prob}(X \in \mathcal{S}) = \text{Prob}(X < \pi/2 + \epsilon) - \text{Prob}(X < \pi/2 - \epsilon) = F(\pi/2 + \epsilon) - F(\pi/2 - \epsilon)$, which converges to zero as ϵ shrinks if F is a continuous function. Defining the probability of X falling in an interval resolves the conceptual problems discussed above, because it is easy to imagine the proportion of days with rainfall in an interval converging to a non-zero value as the sample size increases.

The probability on a small interval is

$$\text{Prob}(x - \epsilon < X < x + \epsilon) = F(x + \epsilon) - F(x - \epsilon) \approx 2\epsilon f(x) \tag{1.12}$$

where $f(x)$ is the derivative of $F(x)$ and is called the probability density function (PDF). If $f(x)$ is the PDF of X, then the probability of $X \in [a, b]$ is the area under the density curve between a and b (Figure 1.3),

$$\text{Prob}(a \le X \le b) = F(b) - F(a) = \int_a^b f(x)dx. \tag{1.13}$$

The distributions of random variables are usually defined via the PDF. To ensure that the PDF produces valid probability statements we must have $f(x) \ge 0$ for all x and $\int_{-\infty}^{\infty} f(x)dx = 1$. Because $f(x)$ is not a probability, but rather a function used compute probabilities via integration, the PDF can be greater than one for some x so long as it integrates to one.

The formulas for the mean and variance of a continuous random variable resemble those for a discrete random variable except that the sum over the

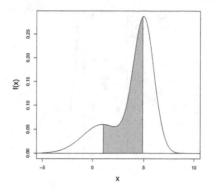

FIGURE 1.3
Computing probabilities using a PDF. The curve is a PDF $f(x)$ and the shaded area is $\text{Prob}(1 < X < 5) = \int_1^5 f(x)dx$.

PMF is replaced with an integral over the PDF,

$$
\begin{aligned}
\text{E}(X) &= \int_{-\infty}^{\infty} x f(x)dx \\
\text{Var}(X) &= \text{E}\{[X - E(X)]^2\} = \int_{-\infty}^{\infty} [x - E(X)]^2 f(x)dx.
\end{aligned}
$$

Another summary that is defined for both discrete and continuous random variables (but is much easier to define in the continuous case) is the quantile function $Q(\tau)$. For $\tau \in [0,1]$, $Q(\tau)$ is the solution to

$$\text{Prob}[X \leq Q(\tau)] = F[Q(\tau)] = \tau. \tag{1.14}$$

That is, $Q(\tau)$ is the value so that the probability of X being no larger than $Q(\tau)$ is τ. The quantile function is the inverse of the distribution function, $Q(\tau) = F^{-1}(\tau)$, and gives the median $Q(0.5)$ and a $(1 - \alpha)\%$ equal-tailed interval $[Q(\alpha/2), Q(1-\alpha/2)]$ so that $\text{Prob}[Q(\alpha/2) \leq X \leq Q(1-\alpha/2)] = 1-\alpha$.

Gaussian: As with discrete data, parametric models are typically assumed for continuous data and practitioners must be familiar with several parametric families. The most common parametric family with support $\mathcal{S} = (-\infty, \infty)$ is the normal (Gaussian) family. The normal distribution has two parameters, the mean $\text{E}(X|\boldsymbol{\theta}) = \mu$ and variance $\text{Var}(X|\boldsymbol{\theta}) = \sigma^2$, and the familiar bell-shaped PDF

$$f(x|\boldsymbol{\theta}) = \frac{1}{\sqrt{2\pi}\sigma} \exp\left[-\frac{1}{2\sigma^2}(x - \mu)^2\right], \tag{1.15}$$

where $\boldsymbol{\theta} = (\mu, \sigma^2)$. The Gaussian distribution is famous because of the central

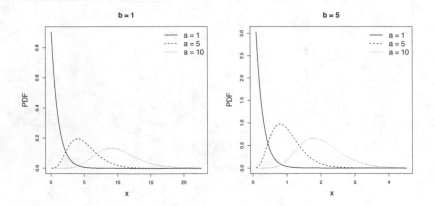

FIGURE 1.4
Plots of the gamma PDF. Plots of the gamma density function $f(x|\boldsymbol{\theta}) = \frac{b^a}{\Gamma(a)}x^a \exp(-bx)$ for several combinations of a and b.

limit theorem (CLT). The CLT applies to the distribution of the sample mean $\bar{X}_n = \frac{\sum_{i=1}^{n} X_i}{n}$, where $X_1, ..., X_n$ are independent samples from some distribution $f(x)$. The CLT says that under fairly general conditions, for large n the distribution of $\bar{X}_n$ is approximately normal even if $f(x)$ is not. Therefore, the Gaussian distribution is a natural model for data that are defined as averages, but can be used for other data as well. Appendix A.1 gives other continuous distributions with $\mathcal{S} = (-\infty, \infty)$ including the double exponential and student-t distributions.

   Gamma: The gamma distribution has $\mathcal{S} = [0, \infty)$. The PDF is

$$f(x|\boldsymbol{\theta}) = \begin{cases} \frac{b^a}{\Gamma(a)}x^a \exp(-bx) & x \geq 0 \\ 0 & x < 0 \end{cases} \qquad (1.16)$$

where Γ is the gamma function and $a > 0$ and $b > 0$ are the two parameters in $\boldsymbol{\theta} = (a, b)$. Beware that the gamma PDF is also written with b in the denominator of the exponential function, but we use the parameterization above. Under the parameterization in (1.16) the mean is a/b and the variance is a/b^2. As shown in Figure 1.4, a is the shape parameter and b is the scale. Setting $a = 1$ gives the exponential distribution with PDF $f(x|\boldsymbol{\theta}) = b\exp(-bx)$ which decays exponentially from the origin and large a gives approximately a normal distribution. Varying b does not change the shape of the PDF but only affects its spread. Appendix A.1 gives other continuous distributions with $\mathcal{S} = [0, \infty)$ including the inverse-gamma distribution.

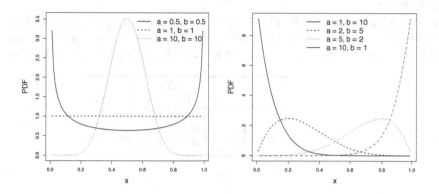

FIGURE 1.5
Plots of the beta PDF. Plots of the beta density function $f(x|\theta) = \frac{\Gamma(a,b)}{\Gamma(a)\Gamma(b)} x^{a-1}(1-x)^{b-1}$ for several a and b.

Beta: The beta distribution has $\mathcal{S} = [0,1]$ and PDF

$$f(x|\theta) = \begin{cases} \frac{\Gamma(a,b)}{\Gamma(a)\Gamma(b)} x^{a-1}(1-x)^{b-1} & x \in [0,1] \\ 0 & x < 0 \text{ or } x > 1, \end{cases} \qquad (1.17)$$

where $a > 0$ and $b > 0$ are the two parameters in $\theta = (a, b)$. As shown in Figure 1.5, the beta distribution is quite flexible and can be left-skewed, right-skewed, or symmetric. The beta distribution also includes the uniform distribution $f(x) = 1$ for $x \in [0,1]$ by setting $a = b = 1$.

1.1.2 Multivariate distributions

Most statistical analyses involve multiple variables with the objective of studying relationships between variables. To model relationships between variables we need multivariate extensions of mass and density functions. Let $X_1, ..., X_p$ be p random variables, $\mathcal{S}_j$ be the support of X_j so that $X_j \in \mathcal{S}_j$, and $\mathbf{X} = (X_1, ..., X_p)$ be the random vector. Table 1.1 describes the joint distribution between the $p = 2$ variables: X_1 indicates that the patient has primary health outcome and X_2 indicates the patient has a side effect. If all variables are discrete, then the joint PMF is

$$\text{Prob}(X_1 = x_1, ..., X_p = x_p) = f(x_1, ..., x_p). \qquad (1.18)$$

To be a valid PMF, f must be non-negative $f(x_1, ..., x_p) \geq 0$ and sum to one $\sum_{x_1 \in \mathcal{S}_1}, ..., \sum_{x_p \in \mathcal{S}_1} f(x_1, ..., x_p) = 1$.
As in the univariate case, probabilities for continuous random variables are

TABLE 1.1
Hypothetical joint PMF. This PMF $f(x_1, x_2)$ gives the probabilities that a patient has a positive ($X_1 = 1$) or negative ($X_1 = 0$) primary health outcome and the patient having ($X_2 = 1$) or not having ($X_2 = 0$) a negative side effect.

		X_2		
		0	1	$f_1(x_1)$
X_1	0	0.06	0.14	0.20
	1	0.24	0.56	0.80
	$f_2(x_2)$	0.30	0.70	

computed indirectly via the PDF $f(x_1, ..., x_p)$. In the univariate case, probabilities are computed as the area under the density curve. For $p > 1$, probabilities are computed as the volume under the density surface. For example, for $p = 2$ random variables, the probability of $a_1 < X_1 < b_1$ and $a_2 < X_2 < b_2$ is

$$\int_{a_1}^{b_1} \int_{a_2}^{b_2} f(x_1, x_2) dx_1 dx_2. \tag{1.19}$$

This gives the probability of the random vector $\mathbf{X} = (X_1, X_2)$ lying in the rectangle defined by the endpoints a_1, b_1, a_2, and b_2. In general, the probability of the random vector $\mathbf{X}$ falling in region $\mathcal{A}$ is the p-dimensional integral $\int_{\mathcal{A}} f(x_1, ..., x_p) dx_1, ..., dx_p$. As an example, consider the bivariate PDF on the unit square with $f(x_1, x_2) = 1$ for $\mathbf{X}$ with $x_1 \in [0, 1]$ and $x_2 \in [0, 1]$ and $f(x_1, x_2) = 0$ otherwise. Then $\text{Prob}(X_1 < .5, X_2 < .1) = \int_0^{0.5} \int_0^{0.1} f(x_1, x_2) dx_1 dx_2 = 0.05$.

1.1.3 Marginal and conditional distributions

The marginal and conditional distributions that follow from a multivariate distribution are key to a Bayesian analysis. To introduce these concepts we assume discrete random variables, but extensions to the continuous case are straightforward by replacing sums with integrals. Further, we assume a bivariate PMF with $p = 2$. Again, extensions to high dimensions are conceptually straightforward by replacing sums over one or two dimensions with sums over $p - 1$ or p dimensions.

The marginal distribution of X_j is simply the distribution of X_j if we consider only a univariate analysis of X_j and disregard the other variable. Denote $f_j(x_j) = \text{Prob}(X_j = x_j)$ as the marginal PMF of X_j. The marginal distribution is computed by summing over the other variable in the joint PMF

$$f_1(x_1) = \sum_{x_2} f(x_1, x_2) \quad \text{and} \quad f_2(x_2) = \sum_{x_1} f(x_1, x_2). \tag{1.20}$$

These are referred to as the marginal distributions because in a two-way table such as Table 1.1, the marginal distributions are the row and column totals of the joint PMF that appear along the table's margins.

As with any univariate distribution, the marginal distribution can be summarized with its mean and variance,

$$\mu_j \;=\; E(X_j) = \sum_{x_j \in \mathcal{S}_j} x_j f_j(x_j) = \sum_{x_1} \sum_{x_2} x_j f(x_1, x_2)$$

$$\sigma_j^2 \;=\; \text{Var}(X_j) = \sum_{x_j} (x_j - \mu_j)^2 f_j(x_j) = \sum_{x_1} \sum_{x_2} (x_j - \mu_j)^2 f(x_1, x_2).$$

The marginal mean and variance measure the center and spread of the marginal distributions, respectively, but do not capture the relationship between the two variables. The most common one-number summary of the joint relationship is covariance, defined as

$$\sigma_{12} \;=\; \text{Cov}(X_1, X_2) = E[(X_1 - \mu_1)(X_2 - \mu_2)] \qquad (1.21)$$

$$=\; \sum_{x_1} \sum_{x_2} (x_1 - \mu_1)(x_2 - \mu_2) f(x_1, x_2).$$

The covariance is often hard to interpret because it depends on the scale of both X_1 and X_2, i.e., if we double X_1 we double the covariance. Correlation is a scale-free summary of the joint relationship, $\rho_{12} = \frac{\sigma_{12}}{\sigma_1 \sigma_2}$. In vector notation, the mean of the random vector $\mathbf{X}$ is $E(\mathbf{X}) = (\mu_1, \mu_2)^T$, the covariance matrix is $\text{Cov}(\mathbf{X}) = \begin{bmatrix} \sigma_1^2 & \sigma_{12} \\ \sigma_{12} & \sigma_2^2 \end{bmatrix}$, and the correlation matrix is $\text{Cor}(\mathbf{X}) = \begin{bmatrix} 1 & \rho_{12} \\ \rho_{12} & 1 \end{bmatrix}$. Generalizing for $p > 2$, the mean vector becomes $E(\mathbf{X}) = (\mu_1, ..., \mu_p)^T$ and the covariance matrix becomes the symmetric $p \times p$ matrix with diagonal elements $\sigma_1^2, ..., \sigma_p^2$ and (i, j) off-diagonal element σ_{ij}.

While the marginal distributions sum over columns or rows of a two-way table, the conditional distributions focus only on a single column or row. The conditional distribution of X_1 given that the random variables X_2 is fixed at x_2 is denoted $f_{1|2}(x_1 | X_2 = x_2)$ or simply $f(x_1 | x_2)$. Referring to Table 1.1, the knowledge that $X_2 = x_2$ restricts the domain to a single column of the two-way table. However, the probabilities in a single column do not define a valid PMF because their sum is less than one. We must rescale these probabilities to sum to one by dividing the column total, which we have previously defined as $f_2(x_2)$. Therefore, the general expression for the conditional distributions of $X_1 | X_2 = x_2$, and similarly $X_2 | X_1 = x_1$, is

$$f_{1|2}(x_1 | X_2 = x_2) = \frac{f(x_1, x_2)}{f_2(x_2)} \quad \text{and} \quad f_{2|1}(x_2 | X_1 = x_1) = \frac{f(x_1, x_2)}{f_1(x_1)}. \qquad (1.22)$$

Atlantic hurricanes example: Table 1.2 provides the counts of Atlantic hurricanes that made landfall between 1990 and 2016 tabulated by their intensity category (1–5) and whether they hit the US or elsewhere. Of course, these are only sample proportions and not true probabilities,

TABLE 1.2
Table of the Atlantic hurricanes that made landfall between 1990 and 2016. The counts are tabulated by their maximum Saffir–Simpson intensity category and whether they made landfall in the US or elsewhere. The counts are downloaded from http://www.aoml.noaa.gov/hrd/hurdat/.

(a) Counts

	Category					
	1	2	3	4	5	Total
US	14	13	10	1	1	39
Not US	46	19	20	17	3	105
Total	60	32	30	18	4	144

(b) Sample proportions

	Category					
	1	2	3	4	5	Total
US	0.0972	0.0903	0.0694	0.0069	0.0069	0.2708
Not US	0.3194	0.1319	0.1389	0.1181	0.0208	0.7292
Total	0.4167	0.2222	0.2083	0.1250	0.0278	1.0000

but for this example we treat Table 1.2b as the joint PMF of location, $X_1 \in \{\text{US}, \text{Not US}\}$, and intensity category, $X_2 \in \{1, 2, 3, 4, 5\}$. The marginal distribution of X_1 is given in the final column and is simply the row sums of the joint PMF. The marginal probability of a hurricane making landfall in the US is $\text{Prob}(X_1 = \text{US}) = 0.2708$, which is the proportion calculation as if we had never considered the storms' category. Similarly, the column sums are the marginal probability of intensity averaging over location, e.g., $\text{Prob}(X_2 = 5) = 0.0278$.

The conditional distributions tell us about the relationship between the two variables. For example, the marginal probability of a hurricane reaching category 5 is 0.0278, but given that the storm hits the US, the conditional distribution is slightly lower, $f_{2|1}(5|\text{US}) = \text{Prob}(X_1 = \text{US}, X_2 = 5)/\text{Prob}(X_1 = \text{US}) = 0.0069/0.2708 = 0.0255$. By definition, the conditional probabilities sum to one,

$$f_{2|1}(1|\text{US}) = \frac{0.0972}{0.2708} = 0.3589,\ f_{2|1}(2|\text{US}) = \frac{0.0903}{0.2708} = 0.3334$$

$$f_{2|1}(3|\text{US}) = \frac{0.0694}{0.2708} = 0.2562,\ f_{2|1}(4|\text{US}) = \frac{0.0069}{0.2708} = 0.0255$$

$$f_{2|1}(5|\text{US}) = \frac{0.0069}{0.2708} = 0.0255.$$

Given that a storm hits the US, the probability of a category 2 or 3 storm increases, while the probability of a category 4 or 5 storm decreases, and so there is a relationship between landfall location and intensity.

Independence example: Consider the joint PMF of the primary health outcome (X_1) and side effect (X_2) in Table 1.1. In this example, the marginal probability of a positive primary health outcome is $\text{Prob}(X_1 = 1) = 0.80$, as is the conditional probability $f_{1|2}(1|X_2 = 1) = 0.56/0.70 = 0.80$ given the patient has the side effect and the conditional probability $f_{1|2}(1|0) = 0.24/0.30 = 0.80$ given the patient does not have the side effect. In other words, both with and without knowledge of side effect status, the probability of a positive health outcome is 0.80, and thus side effect is not informative about the primary health outcome. This is an example of two random variables that are independent.

Generally, X_1 and X_2 are independent if and only if the joint PMF (or PDF for continuous random variables) factors into the product of the marginal distributions,

$$f(x_1, x_2) = f_1(x_1)f_2(x_2). \tag{1.23}$$

From this expression it is clear that if X_1 and X_2 are independent then $f_{1|2}(x_1|X_2 = x_2) = f(x_1, x_2)/f_2(x_2) = f_1(x_1)$, and thus X_2 is not informative about X_1. A special case of joint independence is if all marginal distributions are same, $f_j(x) = f(x)$ for all j and x. In this case, we say that X_1 and X_2 are independent and identically distributed ("iid"), which is denoted $X_j \overset{iid}{\sim} f$. If variables are not independent then they are said to be dependent.

Multinomial: Parametric families are also useful for multivariate distributions. A common parametric family for discrete data is the multinomial, which, as the name implies, is a generalization of the binomial. Consider the case of n independent trials where each trial results in one of p possible outcomes (e.g., $p = 3$ and each result is either win, lose or draw). Let $X_j \in \{0, 1, ..., n\}$ be the number of trails that resulted in outcome j, and $\mathbf{X} = (X_1, ..., X_p)$ be the vector of counts. If we assume that θ_j is the probability of outcome j for each trial, with $\boldsymbol{\theta} = (\theta_1, ..., \theta_p)$ and $\sum_{j=1}^{p} \theta_j = 1$, then $\mathbf{X}|\boldsymbol{\theta} \sim \text{Multinomial}(n, \boldsymbol{\theta})$ with

$$f(x_1, ..., x_p) = \frac{n!}{x_1! \cdot ... \cdot x_p!} \theta_1^{x_1} \cdot ... \cdot \theta_p^{x_p} \tag{1.24}$$

where $n = \sum_{j=1}^{p} x_j$. If there are only $p = 2$ categories then this would be a binomial experiment $X_2 \sim \text{Binomial}(n, \theta_2)$.

Multivariate normal: The multivariate normal distribution is a generalization of the normal distribution to $p > 1$ random variables. For $p = 2$, the bivariate normal has five parameters, $\boldsymbol{\theta} = (\mu_1, \mu_2, \sigma_1^2, \sigma_2^2, \rho)$: a mean parameter for each variable $\text{E}(X_j) = \mu_j$, a variance for each variable $\text{Var}(X_j) = \sigma_j^2 > 0$, and the correlation $\text{Cor}(X_1, X_2) = \rho \in [-1, 1]$. The density function is

$$f(x_1, x_2) = \frac{1}{2\pi\sigma_1\sigma_2\sqrt{1-\rho^2}} \exp\left[-\frac{z_1^2 - 2\rho z_1 z_2 + z_2^2}{2(1-\rho^2)}\right], \tag{1.25}$$

where $z_j = (x_j - \mu_j)/\sigma_j$. As shown in Figure 1.6 the density surface is elliptical with center determined by $\boldsymbol{\mu} = (\mu_1, \mu_2)$ and shape determined by the covariance matrix $\boldsymbol{\Sigma} = \begin{bmatrix} \sigma_1^2 & \sigma_1\sigma_2\rho \\ \sigma_1\sigma_2\rho & \sigma_2^2 \end{bmatrix}$.

A convenient feature of the bivariate normal distribution is that the marginal and conditional distributions are also normal. The marginal distribution of X_j is Gaussian with mean μ_j and variance σ_j^2. The conditional distribution, shown in Figure 1.7, is

$$X_2|X_1 = x_1 \sim \text{Normal}\left[\mu_2 + \rho\frac{\sigma_2}{\sigma_1}(x_1 - \mu_1), (1 - \rho^2)\sigma_2^2\right]. \tag{1.26}$$

If $\rho = 0$ then the conditional distribution is the marginal distribution, as expected. If $\rho > 0$ ($\rho < 0$) then the conditional mean increases (decreases) with x_1. Also, the conditional variance $(1 - \rho^2)\sigma_2^2$ is less than the marginal variance σ_2^2, especially for ρ near -1 or 1, and so conditioning on X_1 reduces uncertainty in X_2 when there is strong correlation.

The multivariate normal PDF for $p > 2$ is most concisely written using matrix notation. The multivariate normal PDF for the random vector $\mathbf{X}$ with mean vector $\boldsymbol{\mu}$ and covariance matrix $\boldsymbol{\Sigma}$ is

$$f(\mathbf{X}) = (2\pi)^{-p/2}|\boldsymbol{\Sigma}|^{-1/2}\exp\left[-\frac{1}{2}(\mathbf{X} - \boldsymbol{\mu})^T\boldsymbol{\Sigma}^{-1}(\mathbf{X} - \boldsymbol{\mu})\right] \tag{1.27}$$

where $|\mathbf{A}|$, $\mathbf{A}^T$ and $\mathbf{A}^{-1}$ are the determinant, transpose and inverse, respectively, of the matrix $\mathbf{A}$. From this expression it is clear that the contours of the log PDF are elliptical. All conditional and marginal distributions are normal, as are all linear combinations $\sum_{j=1}^{p} w_j X_j$ for any $w_1, ..., w_p$.

1.2   Bayes' rule

As the name implies, Bayes' rule (or Bayes' theorem) is fundamental to Bayesian statistics. However, this rule is a general result from probability and follows naturally from the definition of a conditional distribution. Consider two random variables X_1 and X_2 with joint PMF (or PDF as the result holds for both discrete and continuous data) density function $f(x_1, x_2)$. The definition of a conditional distribution gives $f(x_1|x_2) = f(x_2, x_1)/f(x_2)$ and $f(x_2|x_1) = f(x_1, x_2)f(x_1)$. Combining these two expressions gives Bayes' rule

$$f(x_2|x_1) = \frac{f(x_1|x_2)f(x_2)}{f(x_1)}. \tag{1.28}$$

This result is useful as a means to reverse conditioning from $X_1|X_2$ to $X_2|X_1$, and also indicates the need to define a joint distribution for this inversion to be valid.

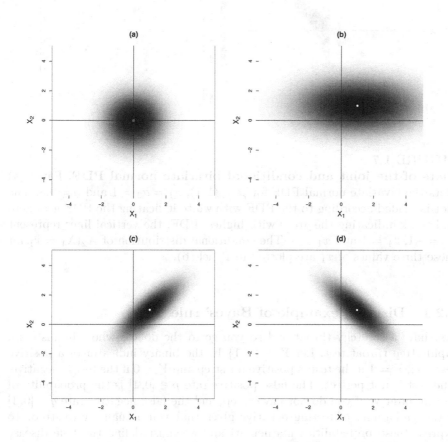

FIGURE 1.6
Plots of the bivariate normal PDF. Panel (a) plots the bivariate normal
PDF for $\mu = (0, 0)$, $\sigma_1 = 1$, $\sigma_2 = 1$ and $\rho = 0$. The other panels modify Panel
(a) as follows: (b) has $\mu = (1, 1)$ and $\sigma_1 = 2$, (c) has $\mu = (1, 1)$ and $\rho = 0.8$,
and (d) has $\mu = (1, 1)$ and $\rho = -0.8$. The plots are shaded according to the
PDF with white indicating the PDF near zero and black indicating the areas
with highest PDF; the white dot is the mean vector μ.

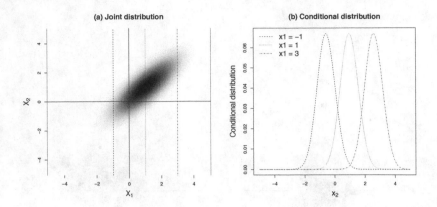

FIGURE 1.7
Plots of the joint and conditional bivariate normal PDF. Panel (a)
plots the bivariate normal PDF for $\mu = (1, 1)$, $\sigma_1 = \sigma_2 = 1$ and $\rho = 0.8$. The
plot is shaded according to the PDF with white indicating the PDF near zero
and black indicating the areas with highest PDF; the vertical lines represent
$x_1 = -1$, $x_1 = 1$ and $x_1 = 3$. The conditional distribution of $X_2|X_1 = x_1$ for
these three values of x_1 are plotted in Panel (b).

1.2.1   Discrete example of Bayes' rule

You have a scratchy throat and so you go to the doctor who administers a
rapid strep throat test. Let $Y \in \{0, 1\}$ be the binary indicator of a positive
test, i.e., $Y = 1$ if the test is positive for strep and $Y = 0$ if the test is negative.
The test is not perfect. The false positive rate $p \in [0, 1]$ is the probability of
testing positive if you do not have strep, and the false negative rate $q \in [0, 1]$
is the probability of testing negative given that you actually have strep. To
express these probabilities mathematically we must define the true disease
status $\theta \in \{0, 1\}$, where $\theta = 1$ if you are truly infected and $\theta = 0$ otherwise.
This unknown variable we hope to estimate is called a parameter. Given these
error rates and the definition of the model parameter, the data distribution
can be written

$$\text{Prob}(Y = 1|\theta = 0) = p \quad \text{and} \quad \text{Prob}(Y = 0|\theta = 1) = q. \tag{1.29}$$

Generally, the PMF (or PDF) of the observed data given the model parameters
is called the likelihood function.

   To formally analyze this problem we must determine which components
should be treated as random variables. Is the test result Y a random variable?
Before the exam, Y is clearly random and (1.29) defines its distribution. This
is aleatoric uncertainty because the results may differ if we repeat the test.
However, after the learning of the test results, Y is determined and you must

decide how to proceed given the value of Y at hand. In this sense, Y is known and no longer random at the analysis stage.

Is the true disease status θ a random variable? Certainly θ is not a random variable in the sense that it changes from second-to-second or minute-to-minute, and so it is reasonable to assume that the true disease status is a fixed quantity for the purpose of this analysis. However, because our test is imperfect we do not know θ. This is epistemic uncertainty because θ is a quantity that we could theoretically know, but at the analysis stage we do not and cannot know θ using only noisy data. Despite our uncertainty about θ, we have to decide what to do next and so it is useful to quantify our uncertainty using the language of probability. If the test is reliable and p and q are both small, then in light of a positive test we might conclude that θ is more likely to be one than zero. But how much more likely? Twice as likely? Three times? In Bayesian statistics we quantify uncertainty about fixed but unknown parameters using probability theory by treating them as random variables. As (1.28) suggests, for formal inversion of conditional probabilities we would need to treat both variables as random.

The probabilities in (1.29) supply the distribution of the test result given disease status, $Y|\theta$. However, we would like to quantify uncertainty in the disease status given the test results, that is, we require the distribution of $\theta|Y$. Since this is the uncertainty distribution after collecting the data this is referred to as the posterior distribution. As discussed above, Bayes' rule can be applied to reverse the order of conditioning,

$$\text{Prob}(\theta = 1 | Y = 1) = \frac{\text{Prob}(Y = 1 | \theta = 1)\text{Prob}(\theta = 1)}{\text{Prob}(Y = 1)}, \quad (1.30)$$

where the marginal probability $\text{Prob}(Y = 1)$ is

$$\sum_{\theta=0}^{1} f(1, \theta) = \text{Prob}(Y = 1 | \theta = 1)\text{Prob}(\theta = 1) + \text{Prob}(Y = 1 | \theta = 0)\text{Prob}(\theta = 0).$$
$$(1.31)$$

To apply Bayes' rule requires specifying the unconditional probability of having strep throat, $\text{Prob}(\theta = 1) = \pi \in [0, 1]$. Since this is the probability of infection before we conduct the test, we refer to this as the prior probability. We can then compute the posterior using Bayes' rule,

$$\text{Prob}(\theta = 1 | Y = 1) = \frac{(1 - q)\pi}{(1 - q)\pi + p(1 - \pi)}. \quad (1.32)$$

To understand this equation consider a few extreme scenarios. Assuming the error rates p and q are not zero or one, if $\pi = 1$ ($\pi = 0$) then the posterior probability of $\theta = 1$ ($\theta = 0$) is one for any value of Y. That is, if we have no prior uncertainty then the imperfect data does not update the prior. Conversely, if the test is perfect and $q = p = 0$ then for any prior $\pi \in (0, 1)$ the posterior probability that θ is Y is one. That is, with perfect data the prior

TABLE 1.3
Strep throat data. Number of patients that are truly positive and tested
positive in the rapid strep throat test data taken from Table 1 of [26].

	Truly positive, test positive	Truly positive, test negative	Truly negative, test positive	Truly negative, test negative
Children	80	38	23	349
Adults	43	10	14	261
Total	123	48	37	610

is irrelevant. Finally, if $p = q = 1/2$, then the test is a random coin flip and
the posterior is the prior $\text{Prob}(\theta = 1|Y) = \pi$.

For a more realistic scenario we use the data in Table 1.3 taken from [26].
We plug in the sample error rates from these data for $p = 37/(37 + 610) =$
0.057 and $q = 48/(48 + 123) = 0.281$. Of course these data represent only
a sample and the sample proportions are not exactly the true error rates,
but for illustration we assume these error rates are correct. Then if we assume
prior probability of disease is $\pi = 0.5$, the posterior probabilities are $\text{Prob}(\theta =$
$1|Y = 0) = 0.230$ and $\text{Prob}(\theta = 1|Y = 1) = 0.927$. Therefore, beginning with
a prior probability of 0.5, a negative test moves the probability down to 0.230
and a positive test increases the probability to 0.927.

Of course, in reality the way individuals process test results is complicated
and subjective. If you have had strep many times before and you went to
the doctor because your current symptoms resemble previous bouts with the
disease, then perhaps your prior is $\pi = 0.8$ and the posterior is $\text{Prob}(\theta =$
$1|Y = 1) = 0.981$. On the other hand, if you went to the doctor only at the
urging of your friend and your prior probability is $\pi = 0.2$, then $\text{Prob}(\theta =$
$1|Y = 1) = 0.759$.

This simple example illustrates a basic Bayesian analysis. The objective
is to compute the posterior distribution of the unknown parameters θ. The
posterior has two ingredients: the likelihood of the data given the parameters
and the prior distribution. Selection of these two distributions is thus largely
the focus of the remainder of this book.

1.2.2 Continuous example of Bayes' rule

Let $\theta \in [0, 1]$ be the proportion of the population in a county that has health
insurance. It is known that the proportion varies across counties following a
$\text{Beta}(a, b)$ distribution and so the prior is $\theta \sim \text{Beta}(a, b)$. We take a sample
of size $n = 20$ from your county and assume that the number of respondents
with insurance, $Y \in \{0, 1, ..., n\}$, is distributed as $Y|\theta \sim \text{Binomial}(n, \theta)$. Joint

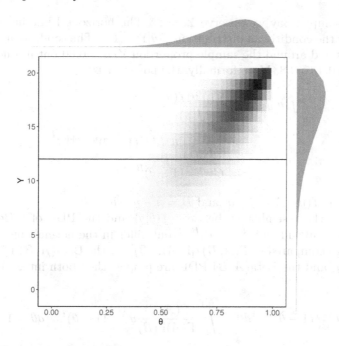

FIGURE 1.8
Joint distribution for the beta-binomial example. Plot of $f(\theta, y)$ for the example with $\theta \sim \text{Beta}(8, 2)$ and $Y|\theta \sim \text{Binomial}(20, \theta)$. The marginal distributions $f(\theta)$ (top) and $f(y)$ (right) are plotted in the margins. The horizontal line is the $Y = 12$ line.

probabilities for θ and Y can be computed from

$$
\begin{aligned}
f(\theta, y) &= f(y|\theta)f(\theta) \\
&= \left\{ \binom{n}{y} \theta^y (1 - \theta)^{n-y} \right\} \left\{ \frac{\Gamma(a, b)}{\Gamma(a)\Gamma(b)} \theta^{a-1} (1 - \theta)^{b-1} \right\} \\
&= c\theta^{y+a-1}(1 - \theta)^{n-y+b-1}
\end{aligned}
$$

where $c = \binom{n}{y}\Gamma(a, b)/[\Gamma(a)\Gamma(b)]$ is a constant that does not depend on θ. Figure 1.8 plots $f(\theta, y)$ and the marginal distributions for θ and Y. By the way we have defined the problem, the marginal distribution of θ, $f(\theta)$, is a Beta(a, b) PDF, which could also be derived by summing $f(\theta, y)$ over y. The marginal distribution of Y plotted on the right of Figure 1.8 is $f(y) = \int_0^1 f(\theta, y)d\theta$. In this case the marginal distribution of Y follows a beta-binomial distribution, but as we will see this is not needed in the Bayesian analysis.

In this problem we are given the unconditional distribution of the disease rate (prior) and the distribution of the sample given the true proportion (likelihood), and Bayes' rule gives the (posterior) distribution of the true proportion given the sample. Say we observe $Y = 12$. The horizontal line in Figure 1.8 traces over the conditional distribution $f(\theta|Y = 12)$. The conditional distribution is centered around the sample proportion $Y/n = 0.60$ but has non-trivial mass from 0.4 to 0.8. More formally, the posterior is

$$
\begin{aligned}
f(\theta|y) &= \frac{f(y|\theta)f(\theta)}{f(y)} \\
&= \left[\frac{c}{f(y)}\right]\theta^{y+a-1}(1-\theta)^{n-y+b-1} \\
&= C\theta^{A-1}(1-\theta)^{B-1}
\end{aligned}
\tag{1.33}
$$

where $C = c/f(y)$, $A = y + a$, and $B = n - y + b$.

We note the resemblance between $f(\theta|y)$ and the PDF of a Beta(A, B) density. Both include $\theta^{A-1}(1-\theta)^{B-1}$ but differ in the normalizing constant, C for $f(\theta|y)$ compared to $\Gamma(A, B)/[\Gamma(A)\Gamma(B)]$ for the Beta(A, B) PDF. Since both $f(\theta|y)$ and the Beta(A, B) PDF are proper, they both integrate to one, and thus

$$
\int_0^1 C\theta^{A-1}(1-\theta)^{B-1}d\theta = \int_0^1 \frac{\Gamma(A, B)}{\Gamma(A)\Gamma(B)}\theta^{A-1}(1-\theta)^{B-1}d\theta = 1
\tag{1.34}
$$

and so

$$
C\int_0^1 \theta^{A-1}(1-\theta)^{B-1}d\theta = \frac{\Gamma(A, B)}{\Gamma(A)\Gamma(B)}\int_0^1 \theta^{A-1}(1-\theta)^{B-1}d\theta
\tag{1.35}
$$

and thus $C = \Gamma(A, B)/[\Gamma(A)\Gamma(B)]$. Therefore, $f(\theta|y)$ is in fact the Beta(A, B) PDF and $\theta|Y = y \sim$ Beta$(y + a, n - y + b)$.

Dealing with the normalizing constant makes posterior calculations quite tedious. Fortunately this can often be avoided by discarding terms that do not involve the parameter of interest and comparing the remaining terms with known distributions. The derivation above can be simplified to (using "$\propto$" to mean "proportional to")

$$
f(\theta|y) \propto f(y|\theta)f(\theta) \propto \theta^{(y+a)-1}(1-\theta)^{(n-y+b)-1}
$$

and immediately concluding that $\theta|Y = y \sim$ Beta$(y + a, n - y + b)$.

Figure 1.9 plots the posterior distribution for two priors and $Y \in \{0, 5, 10, 15, 20\}$. The plots illustrate how the posterior combines information from the prior and the likelihood. In both plots, the peak of the posterior distribution increases with the observation Y. Comparing the plots shows that the prior also contributes to the posterior. When we observe $Y = 0$ successes, the posterior under the Beta(8,2) prior (left) is pulled from zero to the right by the prior (thick line). Under the Beta(1,1), i.e., the uniform prior, when $Y = 0$ the posterior is concentrated around $\theta = 0$.

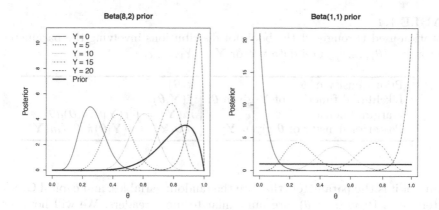

FIGURE 1.9
Posterior distribution for the beta-binomial example. The thick lines
are the beta prior for success probability θ and the thin lines are the posterior
assuming $Y|\theta \sim \text{Binomial}(20, \theta)$ for various values of Y.

1.3 Introduction to Bayesian inference

A parametric statistical analysis models the random process that produced
the data, $\mathbf{Y} = (Y_1, ..., Y_n)$, in terms of fixed but unknown parameters $\boldsymbol{\theta} = (\theta_1, ..., \theta_p)$. The PDF (or PMF) of the data given the parameters, $f(\mathbf{Y}|\boldsymbol{\theta})$, is
called the likelihood function and links the observed data with the unknown
parameters. Statistical inference is concerned with the inverse problem of using
the likelihood function to estimate $\boldsymbol{\theta}$. Of course, if the data are noisy then
we cannot perfectly estimate $\boldsymbol{\theta}$, and a Bayesian quantifies uncertainty about
the unknown parameters by treating them as random variables. Treating $\boldsymbol{\theta}$
as a random variable requires specifying the prior distribution, $\pi(\boldsymbol{\theta})$, which
represents our uncertainty about the parameters before we observe the data.

If we view $\boldsymbol{\theta}$ as a random variable, we can apply Bayes' tule to obtain the
posterior distribution

$$p(\boldsymbol{\theta}|\mathbf{Y}) = \frac{f(\mathbf{Y}|\boldsymbol{\theta})\pi(\boldsymbol{\theta})}{\int f(\mathbf{Y}|\boldsymbol{\theta})\pi(\boldsymbol{\theta})d\boldsymbol{\theta}} \propto f(\mathbf{Y}|\boldsymbol{\theta})\pi(\boldsymbol{\theta}). \tag{1.36}$$

The posterior is proportional to the likelihood times the prior, and quantifies
the uncertainty about the parameters that remain after accounting for prior
knowledge and the new information in the observed data.

Table 1.4 establishes the notation we use throughout for the prior, likelihood and posterior. We will not adhere to the custom (e.g., Section 1.1) that
random variables are capitalized because in a Bayesian analysis more often

TABLE 1.4
Notation used throughout the book for distributions involving the parameter
vector $\boldsymbol{\theta} = (\theta_1, ..., \theta_p)$ and data vector $\mathbf{Y} = (Y_1, ..., Y_n)$.

Prior density of $\boldsymbol{\theta}$:	$\pi(\boldsymbol{\theta})$
Likelihood function of $\mathbf{Y}$ given $\boldsymbol{\theta}$:	$f(\mathbf{Y}\|\boldsymbol{\theta})$
Marginal density of $\mathbf{Y}$:	$m(\mathbf{Y}) = \int f(\mathbf{Y}\|\boldsymbol{\theta})\pi(\boldsymbol{\theta})d\boldsymbol{\theta}$
Posterior density of $\boldsymbol{\theta}$ given $\mathbf{Y}$:	$p(\boldsymbol{\theta}\|\mathbf{Y}) = f(\mathbf{Y}\|\boldsymbol{\theta})\pi(\boldsymbol{\theta})/m(\mathbf{Y})$

than not it is the parameters that are the random variables, and capital Greek
letters, e.g., $\text{Prob}(\Theta = \theta)$, are unfamiliar to most readers. We will however
follow the custom to use bold to represent vectors and matrices. Also, assume
independence unless otherwise noted. For example, if we say "the priors are
$\theta_1 \sim \text{Uniform}(0, 1)$ and $\theta_2 \sim \text{Gamma}(1, 1)$," you should assume that θ_1 and
θ_2 have independent priors.

The Bayesian framework provides a logically consistent framework to use
all available information to quantify uncertainty about model parameters.
However, to apply Bayes' rule requires specifying the prior distribution and the
likelihood function. *How do we pick the prior?* In many cases prior knowledge
from experience, expert opinion or similar studies is available and can be used
to specify an informative prior. It would be a waste to discard this information.
In other cases where prior information is unavailable, then the prior should be
uninformative to reflect this uncertainty. For instance, in the beta-binomial
example in Section 1.2 we might use a uniform prior that puts equal mass on
all possible parameter values. The choice of prior distribution is subjective,
i.e., driven by the analyst's past experience and personal preferences. If a
reader does not agree with your prior then they are unlikely to be persuaded
by your analysis. Therefore, the prior, especially an informative prior, should
be carefully justified, and a sensitivity analysis comparing the posteriors under
different priors should be presented.

How do we pick the likelihood? The likelihood function is the same as in a
classical analysis, e.g., a maximum likelihood analysis. The likelihood function
for multiple linear regression is the product of Gaussian PDFs defined by the
model

$$Y_i|\boldsymbol{\theta} \overset{indep}{\sim} \text{Normal}\left(\beta_0 + \sum_{j=1}^{p} X_{ij}\beta_j, \sigma^2\right) \tag{1.37}$$

where X_{ij} is the value of the jth covariate for the ith observation and
$\boldsymbol{\theta} = (\beta_0, ..., \beta_p, \sigma^2)$ are the unknown parameters. A thoughtful application
of multiple linear regression must consider many questions, including

- Which covariates to include?

- Are the errors Gaussian? Independent? Do they have equal variance?

- Should we include quadratic or interaction effects?

- Should we consider a transformation of the response (e.g., model $\log(Y_i)$)?

- Which observations are outliers? Should we remove them?

- How should we handle the missing observations?

- What p-value threshold should be used to define statistical significance?

As with specifying the prior, these concerns are arguably best resolved using subjective subject-matter knowledge. For example, while there are statistical methods to select covariates (Chapter 5), a more reliable strategy is to ask a subject-matter expert which covariates are the most important to include, at least as an initial list to be refined in the statistical analysis. As another example, it is hard to determine (without a natural ordering as in times series data) whether the observations are independent without consulting someone familiar with the data collection and the study population. Other decisions are made based on visual inspections of the data (such as scatter plots and histograms of the residuals) or ad hoc rules of thumb (threshold on outliers' z-scores or p-values for statistical significance). Therefore, in a typical statistical analysis there many subjective choices to be made, and the choice of prior is far from the most important.

Bayesian statistical methods are often criticized as being subjective. Perhaps an objective analysis that is free from personal preferences or beliefs is an ideal we should strive for (and this is the aim of objective Bayesian methods, see Section 2.3), but it is hard to make the case that non-Bayesian methods are objective, and it can be argued that almost any scientific knowledge and theories are subjective in nature. In an interesting article by Press and Tanur (2001), the authors cite many scientific theories (mainly from physics) where subjectivity played a major role and they concluded *"Subjectivity occurs, and should occur, in the work of scientists; it is not just a factor that plays a minor role that we need to ignore as a flaw..."* and they further added that *"Total objectivity in science is a myth. Good science inevitably involves a mixture of subjective and objective parts."* The Bayesian inferential framework provides a logical foundation to accommodate objective and subjective parts involved in data analysis. Hence, a good scientific practice would be to state upfront all assumptions and then make an effort to validate such assumptions using the current data or preferable future test cases. There is nothing wrong to have a subjective but reasonably flexible model as long as we can exhibit some form of sensitivity analysis when the assumptions of the model are mildly violated.

In addition to explicitly acknowledging subjectivity, another important difference between Bayesian and frequentist (classical) methods is their notion of uncertainty. While a Bayesian considers only the data at hand, a frequentist views uncertainty as arising from repeating the process that generated the data many times. That is, a Bayesian might give a posterior probability that the population mean μ (a parameter) is positive given the data we have observed, whereas a frequentist would give a probability that the sample mean $\bar{Y}$ (a statistic) exceeds a threshold given a specific value of the parameters if we repeated the experiment many times (as is done when computing a p-value).

The frequentist view of uncertainty is well-suited for developing procedures that have desirable error rates when applied broadly. This is reasonable in many settings. For instance, a regulatory agency might want to advocate statistical procedures that ensure only a small proportion of the medications made available to the public have adverse side effects. In some cases however it is hard to see why repeating the sampling is a useful thought experiment. For example, [14] study the relationship between a region's climate and the type of religion that emerged from that region. Assuming the data set consists of the complete list of known cultures, it is hard to imagine repeating the process that led to these data as it would require replaying thousands of years of human history.

Bayesians can and do study the frequentist properties. This is critical to build trust in the methods. If a Bayesian weather forecaster gives the posterior predictive 95% interval every day for a year, but at the end of the year these intervals included the observed temperature only 25% of the time, then the forecaster would lose all credibility. It turns out that Bayesian methods often have desirable frequentist properties, and Chapter 7 examines these properties.

Developing Bayesian methods with good frequentist properties is often called calibrated Bayes (e.g., [52]). According to Rubin [52, 71]: *"The applied statistician should be Bayesian in principle and calibrated to the real world in practice - appropriate frequency calculations help to define such a tie... frequency calculations are useful for making Bayesian statements scientific, scientific in the sense of capable of being shown wrong by empirical test; here the technique is the calibration of Bayesian probabilities to the frequencies of actual events."*

1.4 Summarizing the posterior

The final output of a Bayesian analysis is the posterior distribution of the model parameters. The posterior contains all the relevant information from the data and the prior, and thus all statistical inference should be based on the posterior distribution. However, when there are many parameters, the posterior distribution is a high-dimensional function that is difficult to display graphically, and for complicated statistical models the mathematical form of the posterior may be challenging to work with. In this section, we discuss some methods to summarize a high-dimensional posterior with low-dimensional summaries.

1.4.1 Point estimation

One approach to summarizing the posterior is to use a point estimate, i.e., a single value that represents the best estimate of the parameters given the data and (for a Bayesian analysis) the prior. The posterior mean, median and mode are all sensible choices. Thinking of the Bayesian analysis as a procedure that can be applied to any dataset, the point estimator is an example of an *estimator*, i.e., a function that takes the data as input and returns an estimate of the parameter of interest. Bayesian estimators such as the posterior mean can then be seen as competitors to other estimators such as the sample mean estimator for a population mean or a sample variance for a population variance, or more generally as a competitor to the maximum likelihood estimator. We study the properties of these estimators in Chapter 7.

A common point estimator is the maximum a posteriori (MAP) estimator, defined as the value that maximizes the posterior (i.e., the posterior mode),

$$\hat{\boldsymbol{\theta}}_{MAP} = \arg\max_{\boldsymbol{\theta}} \log[p(\boldsymbol{\theta}|\mathbf{Y})] = \arg\max_{\boldsymbol{\theta}} \log[f(\mathbf{Y}|\boldsymbol{\theta})] + \log[\pi(\boldsymbol{\theta})]. \quad (1.38)$$

The second equality holds because the normalizing constant $m(\mathbf{Y})$ does not depend on the parameters and thus does not affect the optimization.

If the prior is uninformative, i.e., mostly flat as a function of the parameters, then the MAP estimator should be similar to the maximum likelihood estimator (MLE)

$$\hat{\boldsymbol{\theta}}_{MLE} = \arg\max_{\boldsymbol{\theta}} \log[f(\mathbf{Y}|\boldsymbol{\theta})]. \quad (1.39)$$

In fact, this relationship is often used to intuitively justify maximum likelihood estimation. The addition of the log prior $\log[\pi(\boldsymbol{\theta})]$ in (1.38) can be viewed a regularization or penalty term to add stability or prior knowledge.

Point estimators are often useful as fast methods to estimate the parameters for purpose of making predictions. However, a point estimate alone does not quantify uncertainty about the parameters. Sections 1.4.2 and 1.4.3 provide more thorough summaries of the posterior for univariate and multivariate problems, respectively.

1.4.2 Univariate posteriors

A univariate posterior (i.e., from a model with $p = 1$ parameter) is best summarized with a plot because this retains all information about the parameter. Figure 1.10 shows a hypothetical univariate posterior with PDF centered at 0.8 and most of its mass on $\theta > 0.4$.

Point estimators such as the posterior mean or median summarize the center of the posterior, and should be accompanied by a posterior variance or standard deviation to convey uncertainty. The posterior standard deviation resembles a frequentist standard error in that if the posterior is approximately Gaussian then the posterior probability that the parameter is within two posterior standard deviation units of the posterior mean is roughly 0.95. However,

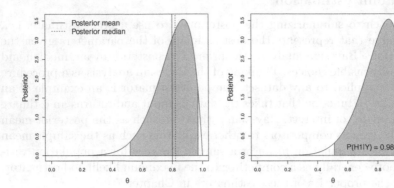

FIGURE 1.10
Summaries of a univariate posterior. The plot on the left gives the posterior mean (solid vertical line), median (dashed vertical line) and 95% equal-tailed interval (shaded area). The plot on the right shades the posterior probability of the hypothesis $H_1 : \theta > 0.5$.

the standard error is the standard deviation of the estimator (e.g., the sample mean) if we repeatedly sample different data sets and compute the estimator for each data set. In contrast, the posterior standard deviation quantifies uncertainty about the parameter given only the single data set under consideration.

The interval $E(\theta|\mathbf{Y}) \pm 2SD(\theta|Y)$ is an example of a credible interval. A $(1 - \alpha)\%$ credible interval is any interval (l, u) so that $\text{Prob}(l < \theta < u|\mathbf{Y}) = 1 - \alpha$. There are infinitely many intervals with this coverage, but the easiest to compute is the equal-tailed interval with l and u set to the $\alpha/2$ and $1 - \alpha/2$ posterior quantiles. An alternative is the highest posterior density interval which searches over all l and u to minimize the interval width $u - l$ while maintaining the appropriate posterior coverage. The HPD thus has the highest average posterior density of all intervals of the form (l, u) that have the nominal posterior probability. As opposed to equal-tailed intervals, the HPD requires an additional optimization step, but this can be computed using the R package `HDInterval`.

Interpreting a posterior credible interval is fairly straightforward. If (l, u) is a posterior 95% interval, this means "given my prior and the observed data, I am 95% sure that θ is between l and u." In a Bayesian analysis we express our subjective uncertainty about unknown parameters by treating them as random variables, and in this subjective sense it is reasonable to assign probabilities to θ. This is in contrast with frequentist confidence intervals which have a more nuanced interpretation. A confidence interval is a procedure that defines an interval for a given data set in a way that ensures the procedure's intervals will include the true value 95% of the time when applied to random datasets.

The posterior distribution can also be used for hypothesis testing. Since the hypotheses are functions of the parameters, we can assign posterior probabilities to each hypothesis. Figure 1.10 (right) plots the posterior probability of the null hypothesis that $\theta < 0.5$ (white) and the posterior probability of the alternative hypothesis that $\theta > 0.5$ (shaded). These probabilities summarize the weight of evidence in support of each hypothesis, and can be used to guide future decisions. Hypothesis testing, and more generally model selection, is discussed in greater detail in Chapter 5.

Summarizing a univariate posterior using R: We have seen that if $Y|\theta \sim \text{Binomial}(n, \theta)$ and $\theta \sim \text{Beta}(a, b)$, then $\theta|Y \sim \text{Beta}(A, B)$, where $A = Y + a$ and $B = n - Y + b$. Listing 1.1 specifies a data set with $Y = 40$ and $n = 100$ and summarizes the posterior using R. Since the posterior is the beta density, the functions dbeta, pbeta and qbeta can be used to summarize the posterior. The posterior median and 95% credible set are 0.401 and (0.309, 0.498).

Monte Carlo sampling: Although univariate posterior distributions are best summarized by a plot, higher dimensional posterior distributions call for other methods such as Monte Carlo (MC) sampling and so we introduce this approach here. MC sampling draws S samples from the posterior, $\theta^{(1)}, ..., \theta^{(S)}$, and uses these samples to approximate the posterior. For example, the posterior mean is approximated using the mean of the S samples, $E(\theta|\mathbf{Y}) \approx \sum_{s=1}^{S} \theta^{(s)}/S$, the posterior 95% credible set is approximated using the 0.025 and 0.975 quantiles of the S samples, etc. Listing 1.1 provides an example of using MC sampling to approximate the posterior mean and 95% credible set.

1.4.3 Multivariate posteriors

Unlike the univariate case, a simple plot of the posterior will not suffice, especially for large p, because plotting in high dimensions is challenging. The typical remedy for this is to marginalize out other parameters and summarize univariate marginal distributions with plots, point estimates, and credible sets, and perhaps plots of a few bivariate marginal distributions (i.e., integrating over the other $p - 2$ parameters) of interest.

Consider the model $Y_i|\mu, \sigma \overset{iid}{\sim} \text{Normal}(\mu, \sigma^2)$ with independent priors $\mu \sim \text{Normal}(0, 100^2)$ and $\sigma \sim \text{Uniform}(0, 10)$ (other priors are discussed in Section 4.1.1). The likelihood

$$f(\mathbf{Y}|\mu, \sigma) \propto \prod_{i=1}^{n} \frac{1}{\sigma} \exp\left[-\frac{(Y_i - \mu)^2}{2\sigma^2}\right] \propto \sigma^{-n} \exp\left[-\frac{\sum_{i=1}^{n}(Y_i - \mu)^2}{2\sigma^2}\right] \quad (1.40)$$

factors as the product of n terms because the observations are assumed to be independent. The prior is $f(\mu, \sigma) = f(\mu)f(\sigma)$ because μ and σ have indepen-

Listing 1.1

Summarizing a univariate posterior in R.

```
1   # Load the data
2   > n <- 100
3   > Y <- 40
4   > a <- 1
5   > b <- 1
6   > A <- Y + a
7   > B <- n - Y + b
8
9   # Define a grid of points for plotting
10  > theta <- seq(0,1,.001)
11
12  # Evaluate the density at these points
13  > pdf <- dbeta(theta,A,B)
14
15  # Plot the posterior density
16  > plot(theta,pdf,type="l",ylab="Posterior",xlab=expression(theta))
17
18  # Posterior mean
19  > A/(A + B)
20  [1] 0.4019608
21
22  # Posterior median (0.5 quantile)
23  > qbeta(0.5,A,B)
24  [1] 0.4013176
25
26  # Posterior probability P(theta<0.5|Y)
27  > pbeta(0.5,A,B)
28  [1] 0.976978
29
30  # Equal-tailed 95% credible interval
31  > qbeta(c(0.025,0.975),A,B)
32  [1] 0.3093085 0.4982559
33
34  # Monte Carlo approximation
35  > S        <- 100000
36  > samples <- rbeta(S,A,B)
37  > mean(samples)
38  [1] 0.402181
39  >   quantile(samples,c(0.025,0.975))
40      2.5%      97.5%
41  0.3092051 0.4973871
```

TABLE 1.5
Bivariate posterior distribution. Summaries of the marginal posterior distributions for the model with $Y_i \overset{iid}{\sim} \text{Normal}(\mu, \sigma^2)$, priors $\mu \sim \text{Normal}(0, 100^2)$ and $\sigma \sim \text{Uniform}(0, 10)$, and $n = 5$ observations $Y_1 = 2.68$, $Y_2 = 1.18$, $Y_3 = -0.97$, $Y_4 = -0.98$, $Y_5 = -1.03$.

Parameter	Posterior mean	Posterior SD	95% credible set
μ	0.17	1.31	(-2.49, 2.83)
σ	2.57	1.37	(1.10, 6.54)

dent priors and since $f(\sigma) = 1/10$ for all $\sigma \in [0, 10]$, the prior becomes

$$\pi(\mu, \sigma) \propto \exp\left(-\frac{\mu^2}{2 \cdot 100^2}\right) \tag{1.41}$$

for $\sigma \in [0, 10]$ and $f(\mu, \sigma) = 0$ otherwise. The posterior is proportional to the likelihood times the prior, and thus

$$p(\mu, \sigma | \mathbf{Y}) \propto \sigma^{-n} \exp\left[-\frac{\sum_{i=1}^{n}(Y_i - \mu)^2}{2\sigma^2}\right] \exp\left(-\frac{\mu^2}{2 \cdot 100^2}\right) \tag{1.42}$$

for $\sigma \in [0, 10]$. Figure 1.11 plots this bivariate posterior assuming there are $n = 5$ observations: $Y_1 = 2.68$, $Y_2 = 1.18$, $Y_3 = -0.97$, $Y_4 = -0.98$, $Y_5 = -1.03$.

The two parameters in Figure 1.11 depend on each other. If $\sigma = 1.5$ (i.e., the conditional distribution traced by the horizontal line at $\sigma = 1.5$ in Figure 1.11) then the posterior of μ concentrates between -1 and 1, whereas if $\sigma = 3$ the posterior of μ spreads from -3 to 3. It is difficult to describe this complex bivariate relationship, so we often summarize the univariate marginal distributions instead. The marginal distributions

$$p(\mu | \mathbf{Y}) = \int_0^{10} p(\mu, \sigma | \mathbf{Y}) d\sigma \quad \text{and} \quad p(\sigma | \mathbf{Y}) = \int_{-\infty}^{\infty} p(\mu, \sigma | \mathbf{Y}) d\mu. \tag{1.43}$$

are plotted on the top (for μ) and right (for σ) of Figure 1.11; they are the row and columns sums of the joint posterior. By integrating over the other parameters, the marginal distribution of a parameter accounts for posterior uncertainty in the remaining parameters. The marginal distributions are usually summarized with point and interval estimates as in Table 1.5.

The marginal distributions and their summaries above were computed by evaluating the joint posterior (1.42) for values of (μ, σ) that form a grid (i.e., pixels in Figure 1.11) and then simply summing over columns or rows of the grid. This is a reasonable approximation for $p = 2$ variables but quickly becomes unfeasible as p increases. Thus, it was only with the advent of more

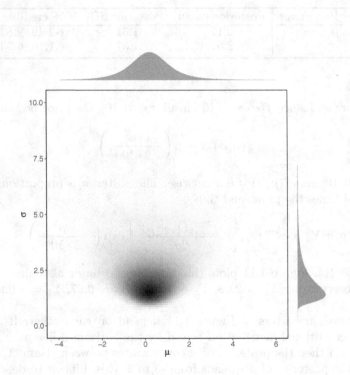

FIGURE 1.11
Bivariate posterior distribution. The bivariate posterior (center) and univariate marginal posteriors (top for μ and right for σ) for the model with
$Y_i \overset{iid}{\sim} \text{Normal}(\mu, \sigma^2)$, priors $\mu \sim \text{Normal}(0, 100^2)$ and $\sigma \sim \text{Uniform}(0, 10)$, and
$n = 5$ observations $Y_1 = 2.68, Y_2 = 1.18, Y_3 = -0.97, Y_4 = -0.98, Y_5 = -1.03$.

efficient computing algorithms in the 1990s that Bayesian statistics became feasible for even medium-sized applications. These exciting computational developments are the subject of Chapter 3.

1.5 The posterior predictive distribution

Often the objective of a statistical analysis is to build a stochastic model that can be used to make predictions of future events or impute missing values. Let Y^* be the future observation we would like to predict. Assuming that the observations are independent given the parameters and that Y^* follows the same model as the observed data, then given θ we have $Y^* \sim f(y|\theta)$ and prediction is straightforward. Unfortunately, we do not know θ exactly, even after observing $\mathbf{Y}$. A remedy for this is to plug in a value of θ, say, the posterior mean $\hat{\theta} = E(\theta|\mathbf{Y})$, and then sample $Y^* \sim f(Y|\hat{\theta})$. However, this ignores uncertainty about the unknown parameters. If the posterior variance of θ is small then its uncertainty is negligible, otherwise a better approach is needed.

For the sake of prediction, the parameters are not of interest themselves, but rather they serve as vehicles to transfer information from the data to the predictive model. We would rather bypass the parameters altogether and simply use the posterior predictive distribution (PPD)

$$Y^*|\mathbf{Y} \sim f^*(Y^*|\mathbf{Y}). \tag{1.44}$$

The PPD is the distribution of a new outcome given the observed data.

In a parametric model, the PPD naturally accounts for uncertainty in the model parameters; this an advantage of the Bayesian framework. The PPD accounts for parametric uncertainty because it can be written

$$f^*(Y^*|\mathbf{Y}) = \int f^*(Y^*, \theta|\mathbf{Y})d\theta = \int f(Y^*|\theta)p(\theta|\mathbf{Y})d\theta, \tag{1.45}$$

where f is the likelihood density (here we assume that the observations are independent given the parameters and so $f(Y^*|\theta) = f(Y^*|\theta, \mathbf{Y})$, and p is the posterior density.

To further illustrate how the PPD accounts for parameteric uncertainty, we consider how to make a sample from the PPD. If we first draw posterior sample $\theta^* \sim p(\theta|\mathbf{Y})$ and then a prediction from the likelihood, $Y^*|\theta^* \sim f(Y|\theta^*)$, then Y^* follows the PPD. A Monte Carlo approximation (Section 3.2) repeats these step many times to approximate the PPD. Unlike the plug-in predictor, each predictive uses a different value of the parameters and thus accurately reflects parametric uncertainty. It can be shown that $\text{Var}(Y^*|\mathbf{Y}) \geq \text{Var}(Y^*|\mathbf{Y}, \theta)$ with equality holding only if there is no posterior uncertainty in the mean of $Y^*|\mathbf{Y}, \theta$.

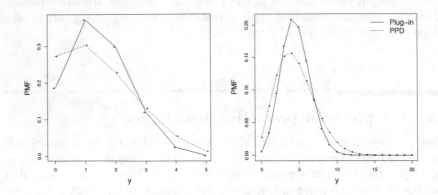

FIGURE 1.12
Posterior predictive distribution for a beta-binomial example. Plots
of the posterior predictive distribution (PPD) from the model $Y|\theta \sim$
Binomial(n, θ) and $\theta \sim$ Beta$(1, 1)$. The "plug-in" PMF is the binomial den-
sity evaluated at the posterior mean $\hat{\theta}$, $f(y|\hat{\theta})$. This is compared with the
full PPD for $Y = 1$ success in $n = 5$ trials (left) and $Y = 4$ successes in
$n = 20$ trials (right). The PMFs are connected by lines for visualization, but
the probabilities are only defined for $y = \{0, 1, ..., n\}$.

As an example, consider the model $Y|\theta \sim$ Binomial(n, θ) and $\theta \sim$
Beta$(1, 1)$. Given the data we have observed (Y and n) we would like to predict the outcome if we repeat the experiment, $Y^* \in \{0, 1, ..., n\}$. The posterior
of θ is $\theta|Y \sim$ Beta$(Y+1, n+1)$ and the posterior mean is $\hat{\theta} = (Y+1)/(n+2)$.
The solid lines in Figure 1.12 show the plug-in prediction $Y^* \sim$ Binomial$(n, \hat{\theta})$
versus the full PPD (Listing 1.2) that accounts for uncertainty in θ (which is
a Beta-Binomial$(n, Y+1, n+1)$ distribution). For both $n = 5$ and $n = 20$,
the PPD is considerably wider than the plug-in predictive distribution, as
expected.

Listing 1.2

Summarizing a posterior predictive distribution (PPD) in R.

```
1  > # Load the data
2  > n <- 5
3  > Y <- 1
4  > a <- 1
5  > b <- 1
6  > A <- Y + a
7  > B <- n - Y + b
8
9  > # Plug-in estimator
10 > theta_hat  <- A/(A+B)
11 > y          <- 0:5
12 > PPD        <- dbinom(y,n,theta_hat)
13 > names(PPD) <- y
14 > round(PPD,2)
15    0    1    2    3    4    5
16 0.19 0.37 0.30 0.12 0.02 0.00
17
18 > # Draws from the PPD, Y_star[i]~Binomial(n,theta_star[i])
19 > S          <- 100000
20 > theta_star <- rbeta(S,A,B)
21 > Y_star     <- rbinom(S,n,theta_star)
22 > PPD        <- table(Y_star)/S
23 > round(PPD,2)
24    0    1    2    3    4    5
25 0.27 0.30 0.23 0.13 0.05 0.01
```

1.6   Exercises

1. If X has support $X \in S = [1, \infty]$, find the constant c (as a function of θ) that makes $f(x) = c \exp(-x/\theta)$ a valid PDF.

2. Assume that $X \sim \text{Uniform}(a, b)$ so the support is $S = [a, b]$ and the PDF is $f(x) = 1/(b - a)$ for any $x \in S$.

   (a) Prove that this is a valid PDF.
   (b) Derive the mean and variance of X.

3. Expert knowledge dictates that a parameter must be positive and that its prior distribution should have the mean 5 and variance 3. Find a prior distribution that satisfies these constraints.

4. X_1 and X_2 have joint PMF

x_1	x_2	$\text{Prob}(X_1 = x_1, X_2 = x_2)$
0	0	0.15
1	0	0.15
2	0	0.15
0	1	0.15
1	1	0.20
2	1	0.20

   (a) Compute the marginal distribution of X_1.
   (b) Compute the marginal distribution of X_2.
   (c) Compute the conditional distribution of $X_1|X_2$.
   (d) Compute the conditional distribution of $X_2|X_1$.
   (e) Are X_1 and X_2 independent? Justify your answer.

5. If (X_1, X_2) is bivariate normal with $E(X_1) = E(X_2) = 0$, $\text{Var}(X_1) = \text{Var}(X_2) = 1$, and $\text{Cor}(X_1, X_2) = \rho$:

   (a) Derive the marginal distribution of X_1.
   (b) Derive the conditional distribution of $X_1|X_2$.

6. Assume (X_1, X_2) have bivariate PDF

$$f(x_1, x_2) = \frac{1}{2\pi} \left(1 + x_1^2 + x_2^2\right)^{-3/2}.$$

   (a) Plot the conditional distribution of $X_1|X_2 = x_2$ for $x_2 \in \{-3, -2, -1, 0, 1, 2, 3\}$ (preferably on the same plot).
   (b) Do X_1 and X_2 appear to be correlated? Justify your answer.
   (c) Do X_1 and X_2 appear to be independent? Justify your answer.

7. According to `insurance.com`, the 2017 auto theft rate was 135 per 10,000 residents in Raleigh, NC compared to 214 per 10,000 residents in Durham/Chapel Hill. Assuming Raleigh's population is twice as large as Durham/Chapel Hill and a car has been stolen somewhere in the triangle (i.e., one of these two areas), what the probability it was stolen in Raleigh?

8. Your daily commute is distributed uniformly between 15 and 20 minutes if there no convention downtown. However, conventions are scheduled for roughly 1 in 4 days, and your commute time is distributed uniformly from 15 to 30 minutes if there is a convention. Let Y be your commute time this morning.

   (a) What is the probability that there was a convention downtown given $Y = 18$?

   (b) What is the probability that there was a convention downtown given $Y = 28$?

9. For this problem pretend we are dealing with a language with a six-word dictionary

$$\{\text{fun, sun, sit, sat, fan, for}\}.$$

An extensive study of literature written in this language reveals that all words are equally likely except that "for" is α times as likely as the other words. Further study reveals that:

   i. Each keystroke is an error with probability θ.

   ii. All letters are equally likely to produce errors.

   iii. Given that a letter is typed incorrectly it is equally likely to be any other letter.

   iv. Errors are independent across letters.

For example, the probability of correctly typing "fun" (or any other word) is $(1 - \theta)^3$, the probability of typing "pun" or "fon" when intending to type is "fun" is $\theta(1-\theta)^2$, and the probability of typing "foo" or "nnn" when intending to type "fun" is $\theta^2(1-\theta)$. Use Bayes' rule to develop a simple spell checker for this language. For each of the typed words "sun", "the", "foo", give the probability that each word in the dictionary was the intended word. Perform this for the parameters below:

   (a) $\alpha = 2$ and $\theta = 0.1$.

   (b) $\alpha = 50$ and $\theta = 0.1$.

   (c) $\alpha = 2$ and $\theta = 0.95$.

Comment on the changes you observe in these three cases.

10. Let $X_1 \sim$ Bernoulli(θ) be the indicator that a tree species occupies a forest and $\theta \in [0, 1]$ denote the prior occupancy probability. The researcher gathers a sample of n trees from the forest and X_2 belong to the species of interest. The model for the data is $X_2|X_1 \sim$ Binomial($n, \lambda X_1$) where $\lambda \in [0, 1]$ the probability of detecting the species given it is present. Give expressions in terms of n, θ and λ for the following joint, marginal and conditional probabilities:

   (a) Prob($X_1 = X_2 = 0$).
   (b) Prob($X_1 = 0$).
   (c) Prob($X_2 = 0$).
   (d) Prob($X_1 = 0|X_2 = 0$).
   (e) Prob($X_2 = 0|X_1 = 0$).
   (f) Prob($X_1 = 0|X_2 = 1$).
   (g) Prob($X_2 = 0|X_1 = 1$).
   (h) Provide intuition for how (d)-(g) change with n, θ and λ.
   (i) Assuming $\theta = 0.5$, $\lambda = 0.1$, and $X_2 = 0$, how large must n be before we can conclude with 95% confidence that the species does not occupy the forest?

11. In a study that uses Bayesian methods to forecast the number of species that will be discovered in future years, [24] report that the number of marine bivalve species discovered each year from 2010-2015 was 64, 13, 33, 18, 30 and 20. Denoting Y_t as the number of species discovered in year t and assuming $Y_t|\lambda \overset{iid}{\sim}$ Poisson(λ) and $\lambda \sim$ Uniform($0, 100$), plot the posterior distribution of λ.

12. Assume that (X, Y) follow the bivariate normal distribution and that both X and Y have marginal mean zero and marginal variance one. We observe six independent and identically distributed data points: (-3.3, -2.6), (0.1, -0.2), (-1.1, -1.5), (2.7, 1.5), (2.0, 1.9) and (-0.4, -0.3). Make a scatter plot of the data and, assuming the correlation parameter ρ has a Uniform($-1, 1$) prior, plot the posterior distribution of ρ.

13. The normalized difference vegetation index (NDVI) is commonly used to classify land cover using remote sensing data. Hypothetically, say that NDVI follows a Beta($25, 10$) distribution for pixels in a rain forest, and a Beta($10, 15$) distribution for pixels in a deforested area now used for agriculture. Assuming about 10% of the rain forest has been deforested, your objective is to build a rule to classify individual pixels as deforested based on their NDVI.

   (a) Plot the PDF of NDVI for forested and deforested pixels, and the marginal distribution of NDVI averaging over categories.

    (b) Give an expression for the probability that a pixel is deforested given its NDVI value, and plot this probability by NDVI.

    (c) You will classify a pixel as deforested if you are at least 90% sure it is deforested. Following this rule, give the range of NDVI that will lead to a pixel being classified as deforested.

14. Let n be the unknown number of customers that visit a store on the day of a sale. The number of customers that make a purchase is $Y|n \sim \text{Binomial}(n, \theta)$ where θ is the known probability of making a purchase given the customer visited the store. The prior is $n \sim \text{Poisson}(5)$. Assuming θ is known and n is the unknown parameter, plot the posterior distribution of n for all combinations of $Y \in \{0, 5, 10\}$ and $\theta \in \{0.2, 0.5\}$ and comment on the effect of Y and θ on the posterior.

15. Last spring your lab planted ten seedlings and two survived the winter. Let θ be the probability that a seedling survives the winter.

    (a) Assuming a uniform prior distribution for θ, compute its posterior mean and standard deviation.

    (b) Assuming the same prior as in (a), compute and compare the equal-tailed and highest density 95% posterior credible intervals.

    (c) If you plant another 10 seedlings next year, what is the posterior predictive probability that at least one will survive the winter?

16. X_1 and X_2 are binary indicators of failure for two parts of a machine. Independent tests have shown that $X_1 \sim \text{Bernoulli}(1/2)$ and $X_2 \sim \text{Bernoulli}(1/3)$. Y_1 and Y_2 are binary indicators of two system failures. We know that $Y_1 = 1$ if both $X_1 = 1$ and $X_2 = 1$ and $Y_1 = 0$ otherwise, and $Y_2 = 0$ if both $X_1 = 0$ and $X_2 = 0$ and $Y_2 = 1$ otherwise. Compute the following probabilities:

    (a) The probability that $X_1 = 1$ and $X_2 = 1$ given $Y_1 = 1$.

    (b) The probability that $X_1 = 1$ and $X_2 = 1$ given $Y_2 = 1$.

    (c) The probability that $X_1 = 1$ given $Y_1 = 1$.

    (d) The probability that $X_1 = 1$ given $Y_2 = 1$.

17. The table below has the overall free throw proportion and results of free throws taken in pressure situations, defined as "clutch" (https://stats.nba.com/), for ten National Basketball Association players (those that received the most votes for the Most Valuable Player Award) for the 2016–2017 season. Since the overall proportion is computed using a large sample size, assume it is fixed and analyze the clutch data for each player separately using Bayesian methods. Assume a uniform prior throughout this problem.

Player	Overall proportion	Clutch makes	Clutch attempts
Russell Westbrook	0.845	64	75
James Harden	0.847	72	95
Kawhi Leonard	0.880	55	63
LeBron James	0.674	27	39
Isaiah Thomas	0.909	75	83
Stephen Curry	0.898	24	26
Giannis Antetokounmpo	0.770	28	41
John Wall	0.801	66	82
Anthony Davis	0.802	40	54
Kevin Durant	0.875	13	16

(a) Describe your model for studying the clutch success probability including the likelihood and prior.

(b) Plot the posteriors of the clutch success probabilities.

(c) Summarize the posteriors in a table.

(d) Do you find evidence that any of the players have a different clutch percentage than overall percentage?

(e) Are the results sensitive to your prior? That is, do small changes in the prior lead to substantial changes in the posterior?

18. In the early twentieth century, it was generally agreed that Hamilton and Madison (ignore Jay for now) wrote 51 and 14 Federalist Papers, respectively. There was dispute over how to attribute 12 other papers between these two authors. In the 51 papers attributed to Hamilton the word "upon" was used 3.24 times per 1,000 words, compared to 0.23 times per 1,000 words in the 14 papers attributed to Madison (for historical perspective on this problem, see [58]).

(a) If the word "upon" is used three times in a disputed text of length 1,000 words and we assume the prior probability 0.5, what is the posterior probability the paper was written by Hamilton?

(b) Give one assumption you are making in (a) that is likely unreasonable. Justify your answer.

(c) In (a), if we changed the number of instances of "upon" to one, do you expect the posterior probability to increase, decrease or stay the same? Why?

(d) In (a), if we changed the text length to 10,000 words and number of instances of "upon" to 30, do you expect the posterior probability to increase, decrease or stay the same? Why?

(e) Let Y be the number of observed number of instances of "upon" in 1,000 words. Compute the posterior probability the paper was written by Hamilton for each $Y \in \{0, 1, ..., 20\}$, plot these posterior probabilities versus Y and give a rule for the number of instances of "upon" needed before the paper should be attributed to Hamilton.

2

From prior information to posterior inference

CONTENTS

One of the most controversial and yet crucial aspects of Bayesian model is the construction of a prior distribution. Often a user is faced with the questions like *Where does the prior distribution come from?* or *What is the true or correct prior distribution?* and so on. There is no concept of the "true, correct or best" prior distribution, but rather the prior distribution can be viewed as an initialization of a statistical (in this case a Bayesian) inferential procedure that gets updated as the data accrue. The choice of a prior distribution is necessary (as you would need to initiate the inferential machine) but there is no notion of the 'optimal' prior distribution. Choosing a prior distribution is similar in principle to initializing any other sequential procedure (e.g., iterative optimization methods like Newton–Raphson, EM, etc.). The choice of such initialization can be good or bad in the sense of the rate of convergence of the procedure to its final value, but as long as the procedure is guaranteed to converge, the choice of prior does not have a permanent impact. As discussed in Section 7.2, the posterior is guaranteed to converge to the true value under very general conditions on the prior distribution.

In this chapter we discuss several general approaches for selecting prior distributions. We begin with conjugate priors. Conjugate priors lead to simple expressions for the posterior distribution and thus illustrate how prior information affects the Bayesian analysis. Conjugate priors are useful when prior information is available, but can also be used when it is not. We conclude with objective Bayesian priors that attempt to remove the subjectivity in prior selection through conventions to be adopted when prior information is not available.

2.1   Conjugate priors

Conjugate priors are the most convenient choice. A prior and likelihood pair are conjugate if the resulting posterior is a member of the same family of distributions as the prior. Therefore, conjugacy refers to a mathematical relationship between two distributions and not to a deeper theme of the appropriate way to express prior beliefs. For example, if we select a beta prior then Figure 1.5 shows that by changing the hyperparameters (i.e., the parameters that define the prior distribution, in this case a and b) the prior could be concentrated around a single value or spread equally across the unit interval.

Because both the prior and posterior are members of the same family, the update from the prior to the posterior affects only the parameters that index the family. This provides an opportunity to build intuition about Bayesian learning through simple examples. Conjugate priors are not unique. For example, the beta prior is conjugate for both the binomial and negative binomial likelihood and both a gamma prior and a Bernoulli prior (trivially) are conjugate for a Poisson likelihood. Also, conjugate priors are somewhat limited because not all likelihood functions have a known conjugate prior, and most conjugacy pairs are for small examples with only a few parameters. These limitations are abated through hierarchical modeling (Chapter 6) and Gibbs sampling (Chapter 3) which provide a framework to build rich statistical models by layering simple conjugacy pairs.

Below we discuss several conjugate priors and mathematical tools needed to derive the corresponding posteriors. Detailed derivation of many of the posterior distributions are deferred to Appendix A.3, and Appendix A.2 has an abridged list of conjugacy results.

2.1.1   Beta-binomial model for a proportion

Returning to the beta-binomial example in Section 1.2, the data $Y \in \{0, 1, ..., n\}$ is the number of successes in n independent trials each with success probability θ. The likelihood is then $Y|\theta \sim \text{Binomial}(n, \theta)$. Since we are only interested in terms in the likelihood that involve θ, we focus only on its kernel, i.e., the terms that involve θ. The kernel of the binomial PMF is

$$f(Y|\theta) \propto \theta^Y (1 - \theta)^{n-Y}. \tag{2.1}$$

If we view the likelihood as a function of θ, it resembles a beta distribution, and so we might suspect that a beta distribution is the conjugate prior for the binomial likelihood. If we select the prior $\theta \sim \text{Beta}(a, b)$, then (as shown in Section 1.2) this combination of likelihood and prior leads to the posterior distribution

$$\theta|Y \sim \text{Beta}(A, B), \tag{2.2}$$

where the updated parameters are $A = Y + a$ and $B = n - Y + b$. Since both the prior and posterior belong to the beta family of distributions, this is an example of a conjugate prior.

The $\text{Beta}(A, B)$ distribution has mean $A/(A + B)$, and so the prior and posterior means are

$$\text{E}(\theta) = \hat{\theta}_0 = \frac{a}{a+b} \quad \text{and} \quad \text{E}(\theta|Y) = \hat{\theta}_1 = \frac{Y+a}{n+a+b}. \tag{2.3}$$

The prior and posterior means are both estimators of the population proportion, and thus we denote them as $\hat{\theta}_0$ and $\hat{\theta}_1$, respectively. The prior mean $\hat{\theta}_0$ is an estimator of θ before observing the data, and this is updated to $\hat{\theta}_1$ by the observed data. A natural estimator of the population proportion θ is the sample proportion $\hat{\theta} = Y/n$, which is the number of successes divided by the number of trials. Comparing the sample proportion to the posterior mean, the posterior mean adds a to the number of successes in the numerator and $a + b$ to the number of trials in the denominator. Therefore, we can think of a as the prior number of successes, $a + b$ as the prior number of trials, and thus b as the prior number of failures (see Section 2.1.5 for a tie to natural conjugate priors).

Viewing the hyperparameters as the prior number of successes and failures provides a means of balancing the information in the prior with the information in the data. For example, the $\text{Beta}(0.5, 0.5)$ prior in Figure 1.5 has one prior observation and the uniform $\text{Beta}(1, 1)$ prior contributes one prior success and one prior failure. If the prior is meant to reflect prior ignorance then we should select small a and b, and if there is strong prior information that θ is approximately θ_0, then we should select a and b so that $\theta_0 = a/(a+b)$ and $a + b$ is large.

The posterior mean can also be written as

$$\hat{\theta}_1 = (1 - w_n)\hat{\theta}_0 + w_n\hat{\theta} \tag{2.4}$$

where $w_n = n/(n + a + b)$ is the weight given to the sample proportion and $1 - w_n$ is the weight given to the prior mean. This confirms the intuition that

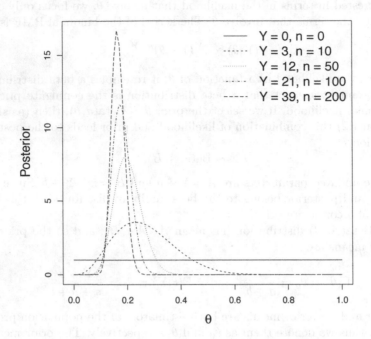

FIGURE 2.1
Posterior distributions from the beta-binomial model. Plot of the posterior of θ from the model $Y|\theta \sim \text{Binomial}(n, \theta)$ and $\theta \sim \text{Beta}(1, 1)$ for various n and Y.

for any prior (a and b), if the sample size n is small the posterior mean is approximately the prior mean, and as the sample size increases the posterior mean becomes closer to the sample proportion. Also, as $a + b \to 0$, $w_n \to 1$, so as we make the prior vague with large variance, the posterior mean coincides with the sample proportion for any sample size n.

Figure 2.1 plots the posterior for various n and Y. This plot is meant to illustrate a sequential analysis. In most cases the entire data set is analyzed in a single analysis, but in some cases data are analyzed as they arrive and a Bayesian analysis provides a framework to analyze data sequentially. In Figure 2.1, before any data are collected θ has a uniform prior. After 10 observations, there are three successes and the posterior concentrates below 0.5. After an additional 40 samples, the posterior centers on the sample proportion 12/50. As data accrues, the posterior converges to $\theta \approx 0.2$ and the posterior variance decreases.

2.1.2   Poisson-gamma model for a rate

The Poisson-gamma conjugate pair is useful for estimating the rate of events per unit of observation effort, denoted θ. For example, an ecologist may survey N acres of a forest and observe $Y \in \{0, 1, ...\}$ individuals of the species of interest, or a company may observe Y employee injuries in N person-hours on the job. For these data, we might assume the model $Y|\theta \sim \text{Poisson}(N\theta)$ where N is known sampling effort and $\theta > 0$ is the unknown event rate. The likelihood is

$$f(Y|\theta) = \frac{\exp(-N\theta)(N\theta)^Y}{Y!} \propto \exp(-N\theta)\theta^Y. \tag{2.5}$$

The kernel of the likelihood resembles a $\text{Gamma}(a, b)$ distribution

$$\pi(\theta) = \frac{b^a}{\Gamma(a)}\exp(-b\theta)\theta^a \propto \exp(-b\theta)\theta^a \tag{2.6}$$

in the sense that θ appears in the PDF raised to a power and in the exponent. Combining the likelihood and prior gives posterior

$$p(\theta|Y) \propto [\exp(-N\theta)\theta^Y] \cdot [\exp(-b\theta)\theta^a] \propto \exp(-B\theta)\theta^A, \tag{2.7}$$

where $A = Y + a$ and $B = N + b$. The posterior of θ is thus $\theta|Y \sim \text{Gamma}(A, B)$, and the gamma prior is conjugate.

A simple estimate of the expected number of events per unit of effort is sample event rate $\hat{\theta} = Y/N$. The mean and variance of a $\text{Gamma}(A, B)$ distribution are A/B and A/B^2, respectively, and so the posterior mean under the Poisson-gamma model is

$$\text{E}(\theta|Y) = \frac{Y + a}{N + b}. \tag{2.8}$$

Therefore, compared to the sample event rate we add a events to the numerator and b units of effort to the denominator. As in the beta-binomial example, the posterior mean can be written as a weighted average of the sample rate and the prior mean,

$$E(\theta|Y) = (1 - w_N)\frac{a}{b} + w_N\frac{Y}{N}, \tag{2.9}$$

where $w_N = N/(N+b)$ is the weight given to the sample rate Y/N and $1 - w_N$ is the weight given to the prior mean a/b. As $b \to 0$, $w_n \to 1$, and so for a vague prior with large variance the posterior mean coincides with the sample rate. A common setting for the hyperparameters is $a = b = \epsilon$ for some small value ϵ, which gives prior mean 1 and large prior variance $1/\epsilon$.

NFL concussions example: Concussions are an increasingly serious concern in the National Football League (NFL). The NFL has 32 teams and each team plays 16 regular-season games per year, for a total of $N = 32 \cdot 16/2 = 256$ games. According to Frontline/PBS (`http://apps.frontline.`

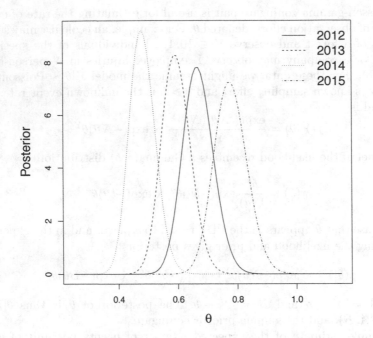

FIGURE 2.2
Posterior distributions for the NFL concussion example. Plot of the
posterior of θ from the model $Y|\theta \sim \text{Gamma}(N\theta)$ and $\theta \sim \text{Gamma}(a,b)$,
where $N = 256$ is the number of games played in an NFL season, Y is the
number of concussions in a year, and $a = b = 0.1$ are the hyperparameters.
The plot below gives the posterior for years 2012–2015, which had 171, 152,
123, and 199 concussions, respectively.

org/concussion-watch) there were $Y_1 = 171$ concussions in 2012, $Y_2 = 152$
concussions in 2013, $Y_3 = 123$ concussions in 2014, and $Y_4 = 199$ concussions
in 2015. Figure 2.2 plots the posterior of the concussion rate for each year
assuming $a = b = 0.1$. Comparing only 2014 with 2015 there does appear to
be some evidence of an increase in the concussion rate per game; for 2014 the
posterior mean and 95% interval are 0.48 (A/B) and (0.40, 0.57) (computed
using the Gamma(A, B) quantile function), compared to 0.78 and (0.67, 0.89)
for 2015. However, 2015 does not appear to be significantly different than 2012
or 2013.

2.1.3   Normal-normal model for a mean

Gaussian responses play a key role in applied and theoretical statistics. The t-test, ANalysis Of VAriance (ANOVA), and linear regression all assume Gaussian responses. We develop these methods in Chapter 4, but here we discuss conjugate priors for the simpler model $Y_i|\mu, \sigma^2 \overset{iid}{\sim} \text{Normal}(\mu, \sigma^2)$ for $i = 1, ..., n$. This model has two parameters, the mean μ and the variance σ^2. In this subsection we assume that σ^2 is fixed and focus on estimating μ; in the following subsection we analyze σ^2 given μ, and Chapter 4 derives the joint posterior of both parameters.

Assuming σ^2 to be fixed, a conjugate prior for the unknown mean μ is $\mu \sim \text{Normal}(\mu_0, \sigma^2/m)$. The prior variance is proportional to σ^2 to express prior uncertainty on the scale of the data. The hyperparameter $m > 0$ controls the strength of the prior, with small m giving large prior variance and vice versa. Appendix A.3 shows that the posterior is

$$\mu|\mathbf{Y} \sim \text{N}\left(w\bar{Y} + (1-w)\mu_0, \frac{\sigma^2}{n+m}\right) \tag{2.10}$$

where $w = n/(n+m) \in [0,1]$ is the weight given to the sample mean $\bar{Y} = \sum_{i=1}^{n} Y_i/n$. Letting $m \to 0$ gives $w_n \to 1$ so the posterior mean coincides with the sample mean as the prior variance increases.

The posterior standard deviation is $\sigma/\sqrt{n+m}$ which is less than the standard error of the sample mean, $\sigma/\sqrt{n}$. The prior with hyperparameter m reduces the posterior standard deviation by the same amount as adding an additional m observations, and this stabilizes a Bayesian analysis. For Gaussian data analyses such as ANOVA or regression the prior mean μ_0 is often set to zero. In the normal-normal model with $\mu_0 = 0$ the posterior mean estimator is $\text{E}(\mu|\mathbf{Y}) = w\bar{Y}$. More generally, $\text{E}(\mu|\mathbf{Y}) = w(\bar{Y} - \mu_0) + \mu_0$. This is an example of a *shrinkage* estimator because the sample mean is shrunk towards the prior mean by the shrinkage factor $w \in [0,1]$. As will be discussed in Chapter 4, shrinkage estimators have advantages, particularly in hard problems such as regression with many predictors and/or small sample size.

Blood alcohol concentration (BAC) example: The BAC level is percent of your blood that is concentrated with alcohol. The legal limit for operating a vehicle is BAC ≤ 0.08 in most US states. Let Y be the measured BAC and μ be your true BAC. Of course, the BAC test has error, and the error standard deviation for a sample near the legal limit has been established in (hypothetical) laboratory tests to be $\sigma = 0.01$, so that the likelihood of the data is $Y|\mu \sim \text{Normal}(\mu, 0.01^2)$. Your BAC is measured to be $Y = 0.082$, which is above the legal limit.

Your defense is that you had two drinks, and that the BAC of someone your size after two drinks has been shown to follow a Normal$(0.05, 0.02^2)$ distribution depending on the person's metabolism and the timing of the drinks. Figure 2.3 plots the prior $\mu \sim \text{Normal}(0.05, 0.02^2)$, i.e., with $m = 0.25$. The prior probability that your BAC exceeds 0.08 is 0.067. The posterior distribu-

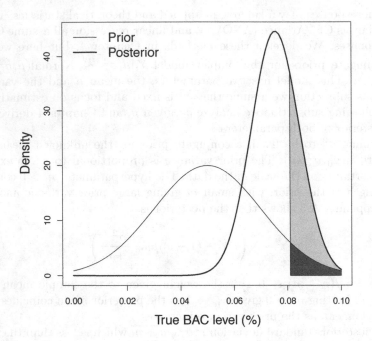

FIGURE 2.3
Posterior distributions for the blood alcohol content example. Plot
of the prior and posterior PDF, with the prior (0.067) and posterior (0.311)
probabilities of exceeding the legal limit of 0.08 shaded.

tion with $n = 1$, $Y = 0.082$, $\sigma = 0.01$ and $m = 0.25$ is

$$\mu|Y \sim \text{Normal}(0.0756, 0.0089^2). \tag{2.11}$$

Therefore, $\text{Prob}(\mu > 0.08|Y) = 0.311$ and so there is considerable uncertainty
about whether your BAC exceeds the legal limit and in fact the posterior odds
that your BAC is below the legal limit,

$$\frac{\text{Prob}(\mu \leq 0.08|Y)}{\text{Prob}(\mu > 0.08|Y)},$$

are greater than two.

2.1.4   Normal-inverse gamma model for a variance

Next we turn to estimating a Gaussian variance assuming the mean is fixed.
As before, the sampling density is $Y_i|\sigma^2 \overset{iid}{\sim} \text{Normal}(\mu, \sigma^2)$ for $i = 1, ..., n$. With

the mean fixed, the likelihood is

$$f(\mathbf{Y}|\sigma^2) \propto \prod_{i=1}^{n} \frac{1}{\sigma} \exp\left[-\frac{(Y_i - \mu)}{2\sigma^2}\right] \propto (\sigma^2)^{-n/2} \exp\left(-\frac{SSE}{2\sigma^2}\right) \qquad (2.12)$$

where $SSE = \sum_{i=1}^{n}(Y_i - \mu)^2$. The likelihood has σ^2 raised to a negative power and σ^2 in the denominator of the exponent. Of the distributions in Appendix A.1 with support $[0, \infty)$, only the inverse gamma PDF has these properties. Taking the prior $\sigma^2 \sim \text{InvGamma}(a, b)$ gives

$$\pi(\sigma^2) \propto (\sigma^2)^{-(a+1)} \exp\left(-\frac{b}{\sigma^2}\right). \qquad (2.13)$$

Combining the likelihood and prior gives the posterior

$$p(\sigma^2|\mathbf{Y}) \propto f(\mathbf{Y}|\sigma^2)\pi(\sigma^2) \propto (\sigma^2)^{-(A+1)} \exp\left(-\frac{B}{\sigma^2}\right), \qquad (2.14)$$

where $A = n/2 + a$ and $B = SSE/2 + b$ are the updated parameters, and therefore

$$\sigma^2|\mathbf{Y} \sim \text{InvGamma}(A, B). \qquad (2.15)$$

The posterior mean (if $n/2 + a > 1$) is

$$E(\sigma^2|\mathbf{Y}) = \frac{SSE/2 + b}{n/2 + a - 1} = \frac{SSE + 2b}{n - 1 + 2a - 1}. \qquad (2.16)$$

Therefore, if we take the hyperparameters to be $a = 1/2 + m/2$ and $b = \epsilon/2$ for small m and ϵ, then the posterior-mean estimator is $E(\sigma^2|\mathbf{Y}) = (SSE + \epsilon)/(n - 1 + m)$ and compared to the usual sample variance $SSE/(n - 1)$ the small values ϵ and m are added to the numerator and denominator, respectively. In this sense, the prior adds an additional m degrees of freedom for estimating the variance, which can stabilize the estimator if n is small.

Conjugate prior for a precision: We have introduced the Gaussian distribution as having two parameters, the mean and the variance. However, it can also be parameterized in terms of its mean and precision (inverse variance), $\tau = 1/\sigma^2$. In particular, the JAGS package used in this book employs this parameterization. In this parameterization, $Y|\mu, \tau \sim \text{Normal}(\mu, \tau)$ has PDF

$$f(Y|\mu, \tau) = \frac{\tau^{1/2}}{\sqrt{2\pi}} \exp\left[-\frac{\tau}{2}(Y - \mu)^2\right]. \qquad (2.17)$$

This parameterization makes derivations and computations slightly easier, especially for the multivariate normal distribution where using a precision matrix (inverse covariance matrix) avoids some matrix inversions.

Not surprisingly, the conjugate prior for the precision is the gamma family. If $Y_i \overset{iid}{\sim} \text{Normal}(\mu, \tau)$ then the likelihood is

$$f(\mathbf{Y}|\tau) \propto \prod_{i=1}^{n} \tau^{1/2} \exp\left[-\frac{\tau}{2}(Y_i - \mu)^2\right] \propto \tau^{n/2} \exp\left(-\frac{\tau}{2}SSE\right), \qquad (2.18)$$

and the Gamma(a, b) prior is $\pi(\tau) \propto \tau^{a-1} \exp(-\tau b)$. Combining the likelihood and prior gives

$$p(\tau|\mathbf{Y}) \propto f(\mathbf{Y}|\tau)\pi(\tau) \propto \tau^{A-1} \exp(-\tau B), \tag{2.19}$$

where $A = n/2 + a$ and $B = SSE/2 + b$ are the updated parameters. Therefore,

$$\tau|\mathbf{Y} \sim \text{Gamma}(A, B). \tag{2.20}$$

The InvGamma(a, b) prior for the variance and the Gamma(a, b) prior for the precision give the exact same posterior distribution. That is, if we use the InvGamma(a, b) prior for the variance and then convert this to obtain the posterior of $1/\sigma^2$, the results are identical as if we have conducted the analysis with a Gamma(a, b) prior for the precision. Throughout the book we use the mean-variance parameterization except for cases involving JAGS code when we adopt their mean-precision parameterization.

2.1.5 Natural conjugate priors

We have thus far catalogued a series of popular sampling distributions and corresponding choices of the conjugate family of priors. Is there a natural way of constructing a class of conjugate priors given a specific sampling density $f(y|\theta)$? It turns out for many of these familiar choices of sampling densities (e.g., exponential family) we can construct a class of conjugate priors, and priors constructed in this manner are called natural conjugate priors.

Let $Y_i \overset{iid}{\sim} f(y|\theta)$ for $i = 1, \ldots, n$ and consider a class of priors defined by

$$\pi(\theta|y_1^0, \ldots, y_m^0, m) \propto \prod_{j=1}^{m} f(y_j^0|\theta) \tag{2.21}$$

where y_j^0 are some arbitrary fixed values in the support of the sampling distribution for $j = 1, \ldots, m$ and $m \geq 1$ is a fixed integer. The pseudo-observations y_j^0 can be seen as the hyperparameters of the prior distribution and such a prior is a proper distribution if there exists m such that $\int \prod_{j=1}^{m} f(y_j^0|\theta) \, d\theta < \infty$. To see that the prior defined in (2.21) is indeed conjugate notice that

$$p(\theta|\mathbf{Y}) \propto \pi(\theta|y_1^0, \ldots, y_m^0, m) \prod_{i=1}^{n} f(Y_i|\theta) \propto \prod_{j=1}^{M} f(y_j^*|\theta) \tag{2.22}$$

where $M = m + n$ and $y_j^* = y_j^0$ for $j = 1, \ldots, m$ and $y_j^* = Y_{j-m}$ for $j = m+1, \ldots, m+n$. Thus, the posterior distribution belongs to the same class of distributions as in (2.21).

Below we revisit some previous examples using this method of creating natural conjugate priors.

Bernoulli trials (Section 2.1.1): When $Y|\theta \sim$ Bernoulli(θ), we have

$f(y|\theta) \propto \theta^y(1-\theta)^{1-y}$ and so the natural conjugate prior with the first s_0 pseudo observations equal $y_j^0 = 1$ and the remaining $m - s_0$ pseudo observations set to $y_j^0 = 0$ gives the prior $\pi(\theta|y_1^0, \ldots, y_m^0, m) \propto \theta^{s_0}(1-\theta)^{m-s_0}$, that is, $\theta \sim \text{Beta}(s_0+1, m-s_0+1)$. This beta prior is restricted to have integer-valued hyperparameters, but once we see the form we can relax the assumption that m and s_0 are integer valued.

Poisson counts (Section 2.1.2): When $Y|\theta \sim \text{Poisson}(\theta)$, we have $f(y|\theta) \propto \theta^y e^{-\theta}$ and so the natural conjugate prior by using equation (2.21) is given by $\pi(\theta|y_1^0, \ldots, y_m^0, m) \propto \theta^{s_0} e^{-m\theta}$, where $s_0 = \sum_{j=1}^m y_j^0$ and this naturally leads to a gamma prior distribution $\theta \sim \text{Gamma}(s_0 + 1, m)$. Therefore, we can view the prior as consisting of m pseudo observations with sample rate s_0/m. Again, the restriction that m is an integer can be relaxed this once it is revealed that the prior is from the gamma family.

Normal distribution with fixed variance (Section 2.1.3): Assuming $Y|\theta \sim \text{Normal}(\mu, 1)$, we have $f(y|\mu) \propto \exp\{-(y-\mu)^2/(2\sigma^2)\}$ and so the natural conjugate prior is $\pi(\mu|y_1^0, \ldots, y_m^0, m) \propto \exp\{-\sum_{j=1}^m (y_j^0 - \mu)^2/(2\sigma^2)\} \propto \exp\{-m(\mu - \bar{y}^0)^2/(2\sigma^2)\}$, where $\bar{y}^0 = \sum_{j=1}^m y_j^0/m$ and this naturally leads to the prior $\theta \sim \text{Normal}(\bar{y}^0, \sigma^2/m)$. Therefore, the prior can be viewed as consisting of m pseudo observations with mean $\bar{y}^0$.

This systematic way of obtaining a conjugate prior takes away the mystery of first guessing the form of the conjugate prior and then verifying its conjugacy. The procedure works well when faced with problems that do not have a familiar likelihood and works even when we have vector-valued parameters.

2.1.6 Normal-normal model for a mean vector

In Bayesian linear regression the parameter of interest is the vector of regression coefficients. This is discussed extensively in Chapter 4, but here we provide the conjugacy relationship that underlies the regression analysis. Although still a bit cumbersome, linear regression notation is far more concise using matrices. Say the n-vector $\mathbf{Y}$ is multivariate normal

$$\mathbf{Y}|\boldsymbol{\beta} \sim \text{Normal}(\mathbf{X}\boldsymbol{\beta}, \boldsymbol{\Sigma}). \tag{2.23}$$

The mean of $\mathbf{Y}$ is decomposed as the known $n \times p$ matrix $\mathbf{X}$ and unknown p-vector $\boldsymbol{\beta}$, and $\boldsymbol{\Sigma}$ is the $n \times n$ covariance matrix. The prior for $\boldsymbol{\beta}$ is multivariate normal with mean $\boldsymbol{\mu}$ and $p \times p$ covariance matrix $\boldsymbol{\Omega}$,

$$\boldsymbol{\beta} \sim \text{Normal}(\boldsymbol{\mu}, \boldsymbol{\Omega}). \tag{2.24}$$

As shown in Appendix A.3, the posterior of $\boldsymbol{\beta}$ is multivariate normal

$$\boldsymbol{\beta}|\mathbf{Y} \sim \text{Normal}\left[\boldsymbol{\Sigma}_\beta(\mathbf{X}^T\boldsymbol{\Sigma}^{-1}\mathbf{Y} + \boldsymbol{\Omega}^{-1}\boldsymbol{\mu}), \boldsymbol{\Sigma}_\beta\right], \tag{2.25}$$

where $\boldsymbol{\Sigma}_\beta = (\mathbf{X}'\boldsymbol{\Sigma}^{-1}\mathbf{X} + \boldsymbol{\Omega}^{-1})^{-1}$. In standard linear regression the errors are assumed to be independent and identically distributed and thus the covariance is proportional to the identity matrix, $\boldsymbol{\Sigma} = \sigma^2 \mathbf{I}_n$. In this case, if the prior is uninformative with $\boldsymbol{\Omega}^{-1} \approx 0$, then the posterior mean reduces to the familiar least squares estimator $(\mathbf{X}^T \mathbf{X})^{-1} \mathbf{X}^T \mathbf{Y}$ and the posterior covariance reduces to the covariance of the sampling distribution of the least squares estimator, $\sigma^2 (\mathbf{X}^T \mathbf{X})^{-1}$.

2.1.7   Normal-inverse Wishart model for a covariance matrix

Say $\mathbf{Y}_1, ..., \mathbf{Y}_n$ are vectors of length p that are independently distributed as multivariate normal with known mean vectors $\boldsymbol{\mu}_i$ and $p \times p$ unknown covariance matrix $\boldsymbol{\Sigma}$. The conjugate prior for the covariance matrix is the inverse Wishart prior. The inverse Wishart family's support is symmetric positive definite matrices (i.e., covariance matrices), and reduces to the inverse gamma family if $p = 1$. The inverse Wishart prior with degrees of freedom $\nu > p - 1$ and $p \times p$ symmetric and positive definite scale matrix $\mathbf{R}$ has PDF

$$\pi(\boldsymbol{\Sigma}) \propto |\boldsymbol{\Sigma}|^{-(\nu+p+1)/2} \exp\left[-\frac{1}{2}\text{Trace}(\boldsymbol{\Sigma}^{-1}\mathbf{R})\right] \tag{2.26}$$

and mean $E(\boldsymbol{\Sigma}) = \mathbf{R}/(\nu - p - 1)$ assuming $\nu > p + 1$. The prior concentration around $\mathbf{R}/(\nu - p - 1)$ increases with ν, and therefore small ν, say $\nu = p - 1 + \epsilon$ for small $\epsilon > 0$, gives the least informative prior. An interesting special case is when $\nu = p + 1$ and $\mathbf{R}$ is a diagonal matrix. This induces a uniform prior on each off-diagonal element of the correlation matrix corresponding to covariance matrix $\boldsymbol{\Sigma}$.

As shown in Chapter Appendix A.3, the posterior is

$$\boldsymbol{\Sigma}|\mathbf{Y} \sim \text{InvWishart}_p\left(n + \nu, \sum_{i=1}^n (\mathbf{Y}_i - \boldsymbol{\mu}_i)(\mathbf{Y}_i - \boldsymbol{\mu}_i)^T + \mathbf{R}\right). \tag{2.27}$$

The posterior mean is $[\sum_{i=1}^n (\mathbf{Y}_i - \boldsymbol{\mu}_i)(\mathbf{Y}_i - \boldsymbol{\mu}_i)^T + \mathbf{R}]/[n + \nu - p - 1]$. For $\mathbf{R} \approx 0$ and $\nu \approx p + 1$, the posterior mean is approximately $[\sum_{i=1}^n (\mathbf{Y}_i - \boldsymbol{\mu}_i)(\mathbf{Y}_i - \boldsymbol{\mu}_i)^T]/n$, which is the sample covariance matrix assuming the means are known and not replaced by the sample means.

Marathon example: Figure 2.4 plots the data for several of the top female runners in the 2016 Boston Marathon. Let Y_{ij} be the speed (minutes/mile) for runner $i = 1, ..., n$ and mile $j = 1, ..., p = 26$. For this analysis, we have discarded all runners with missing data (for a missing data analysis see Section 6.4), leaving $n = 59$ observations $\mathbf{Y}_i = (Y_{i1}, ..., Y_{ip})^T$. We analyze the covariance of the runners' data to uncover patterns and possibly strategy.

For simplicity we conduct the analysis conditioning on $\mu_{ij} = \bar{Y}_j = \sum_{i=1}^n Y_{ij}/n$, i.e., the sample mean for mile j. For the prior, we take $\nu = p + 1$ and $\mathbf{R} = \mathbf{I}_p/\nu$ so that elements of the correlation matrix have Uniform(-1,1) priors. The code in Listing 2.1 generates S samples from the posterior of $\boldsymbol{\Sigma}$ and uses the samples to approximate the posterior mean. To avoid storage problems, rather than storing all S samples the code simply retains the running

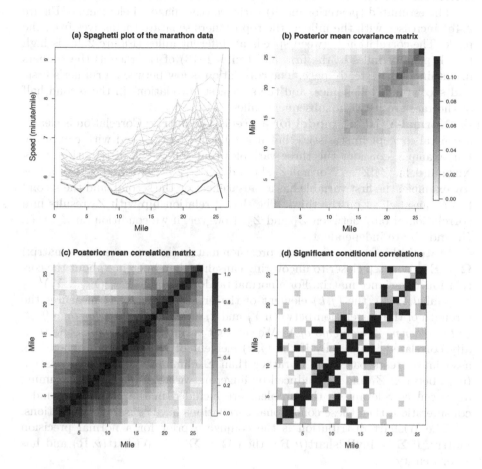

FIGURE 2.4
Covariance analysis of the 2016 Boston Marathon data. Panel (a) plots the minute/mile for each runner (the winner is in black), Panels (b) and (c) show the posterior mean of the covariance and correlation matrices, respectively, and Panel (d) is shaded gray (black) if the 95% (99%) credible interval for the elements of the precision matrix excludes zero.

mean of the samples. However, Monte Carlo sampling produces the full joint distribution of all elements of $\boldsymbol{\Sigma}$. In particular, for each draw from the posterior, Listing 2.1 converts the simulated covariance matrix to the corresponding correlation matrix, and computes the posterior mean of the correlation matrix.

The estimated (posterior mean) variance (the diagonal elements of Figure 2.4b) increases with the mile as the top runners separate themselves from the pack. The correlation between speeds at different miles (Figure 2.4c) is high for all pairs of miles in the first half (miles 1–13) of the race as the runners maintain a fairly steady pace. The correlation is low between a runner's first- and second-half mile times, and the strongest correlations in the second half of the race are between subsequent miles.

Normal-Wishart model for a precision matrix: Correlation is clearly a central concept in statistics, but it should not be confused with causation. For example, consider the three variables that follow the distribution $Z_1 \sim$ Normal$(0, 1)$, $Z_2|Z_1 \sim$ Normal$(Z_1, 1)$ and $Z_3|Z_1, Z_2 \sim$ Normal$(Z_2, 1)$. In this toy example, the first variable has a causal effect on the second, and the second has a causal effect on the third. The shared relationship with Z_2 results in a correlation of 0.57 between Z_1 and Z_3. However, if we condition on Z_2, then Z_1 and Z_3 are independent.

Statistical inference about the precision matrix (inverse covariance matrix) $\boldsymbol{\Omega} = \boldsymbol{\Sigma}^{-1}$ is a step closer to uncovering causality than inference about the correlation/covariance matrix. For a normal random variable $\mathbf{Y} = (Y_1, ..., Y_p)^T \sim$ Normal$(\boldsymbol{\mu}, \boldsymbol{\Omega}^{-1})$, the (j, k) element of the precision matrix $\boldsymbol{\Omega}$ measures the strength of the correlation between Y_j and Y_k *after accounting for the effects of the other $p - 2$ elements of* $\mathbf{Y}$. We say that variables j and k are conditionally correlated if and only if the (j, k) element of $\boldsymbol{\Omega}$ is non-zero. Therefore, association tests based on $\boldsymbol{\Omega}$ rather than $\boldsymbol{\Sigma}$ eliminate spurious correlation (e.g., between Z_1 and Z_3) induced by lingering variables (e.g., Z_2). Assuming all variables relevant to the problem are included in the p variables under consideration, then these conditional correlations have causal interpretations.

The Wishart distribution is the conjugate prior for a normal precision matrix. If $\boldsymbol{\Sigma} \sim$ InvWishart$(\nu, \mathbf{R})$, then $\boldsymbol{\Omega} = \boldsymbol{\Sigma}^{-1} \sim$ Wishart$(\nu, \mathbf{R})$ and has prior density

$$\pi(\boldsymbol{\Omega}) \propto |\boldsymbol{\Omega}|^{(p-\nu-1)/2} \exp\left[-\text{Trace}\left(\boldsymbol{\Omega}\mathbf{R}^{-1}\right)/2\right]. \tag{2.28}$$

Given a sample $\mathbf{Y}_1, ..., \mathbf{Y}_n \sim$ Normal$(\boldsymbol{\mu}, \boldsymbol{\Omega}^{-1})$ and conditioning on the mean vectors $\boldsymbol{\mu}_i$, the posterior is

$$\boldsymbol{\Omega}|\mathbf{Y} \sim \text{Wishart}\left(n + \nu, \left[\sum_{i=1}^{n}(\mathbf{Y}_i - \boldsymbol{\mu}_i)(\mathbf{Y}_i - \boldsymbol{\mu}_i)^T + \mathbf{R}^{-1}\right]^{-1}\right). \tag{2.29}$$

Marathon example: Listing 2.1 computes and Figure 2.4d plots the conditional correlations with credible sets that exclude zero for the Boston Marathon example. Many of the non-zero correlations in Figure 2.4c do not

Listing 2.1
Monte Carlo analysis of the Boston Marathon data.

```
1
2   # Hyperpriors
3
4   nu        <- p+1
5   R         <- diag(p)/(p+1)
6
7   # Process the data
8
9   Ybar      <- colMeans(Y)
10  SSE       <- sweep(Y,2,Ybar,"-")
11  SSE       <- solve(t(SSE)%*%SSE)
12
13  # Monte Carlo settings from the posterior
14
15  S         <- 10000
16  Sigma_mn <- Cor_mn <- Omega_pos <- 0
17
18  # Monte Carlo sampling
19
20  for(s in 1:S){
21    Omega     <- rwish(n+nu,SSE+R)
22    Sigma     <- solve(Omega)
23    Sigma_mn <- Sigma_mn + Sigma/S
24    Cor_mn    <- Cor_mn + cov2cor(Sigma)/S
25    Omega_pos <- Omega_pos + (Omega>0)/S
26  }
27
28  # Evaluate significance of the precision matrix
29
30  Omega_sig <- ifelse(Omega_pos<0.025 | Omega_pos>0.975,
31                    "gray","white")
32  Omega_sig <- ifelse(Omega_pos<0.005 | Omega_pos>0.995,
33                    "black",Omega_sig)
34
35  # Plot some of the results
36
37  library(fields)
38  image.plot(1:p,1:p,Sigma_mn,
39            xlab="Mile",ylab="Mile",
40            main="Posterior mean covariance matrix")
```

have significant conditional correlations. For example, if the times of the first 25 miles are known, then only miles 20 and 25 have strong conditional correlation with the final mile time. Therefore, despite the many strong correlations between the final mile and the other 25 miles, for the purpose of predicting the final mile time given all other mile times, using only miles 20 and 25 may be sufficient.

2.1.8 Mixtures of conjugate priors

A limitation of a conjugate prior is that restricting the prior to a parametric family limits how accurately prior uncertainty can be expressed. For example, the normal distribution is a conjugate prior for a normal mean, but a normal prior must be symmetric and unimodal, which may not describe the available prior information. A mixture of conjugate priors provides a much richer class of priors. A mixture prior is

$$\pi(\boldsymbol{\theta}) = \sum_{j=1}^{J} q_j \pi_j(\boldsymbol{\theta}) \tag{2.30}$$

where $q_1, ..., q_J$ are the mixture probabilities with $q_j \geq 0$ and $\sum_{j=1}^{J} q_j = 1$, and π_j are valid PDFs (or PMFs). The mixture prior can be interpreted as each component corresponding to the opinion of a different expert. That is, with probability q_j you select the prior from expert j, denoted π_j.

The posterior distribution under a mixture prior is

$$p(\boldsymbol{\theta}|\mathbf{Y}) \propto f(\mathbf{Y}|\boldsymbol{\theta})\pi(\boldsymbol{\theta}) \propto \sum_{j=1}^{p} q_j f(\mathbf{Y}|\boldsymbol{\theta})\pi_j(\boldsymbol{\theta}) \propto \sum_{j=1}^{p} Q_j p_j(\boldsymbol{\theta}|\mathbf{Y}), \tag{2.31}$$

where $p_j(\boldsymbol{\theta}|\mathbf{Y}) \propto f(\mathbf{Y}|\boldsymbol{\theta})\pi_j(\boldsymbol{\theta})$ is the posterior under prior π_j and the updated mixture weights Q_j are positive and sum to one. The posterior mixture weights Q_j are not the same is the prior mixture weights q_j, and unfortunately have a fairly complicated form. However, since each mixture component is conjugate, the p_j belong to the same family as the π_j. Therefore, the posterior is also a mixture distribution over the same parametric family as the prior, and thus the prior is conjugate.

ESP example: As discussed in [56], [7] presents the results of several experiments on extrasensory perception (ESP). In one experiment (Experiment 2 in [7]), subjects were asked to say which of two pictures they preferred before being shown the pictures. Unknown to the subject, one of the pictures was a subliminally exposed negative picture, and the subjects avoided this picture in 2,790 of the 5,400 trials (51.7%). Even more dramatically, participants that scored high in stimulus seeking avoided the picture in $Y = 963$ of the $n = 1,800$ trials (53.5%).

The data are then $Y \sim \text{Binomial}(n, \theta)$, and the objective is to test whether $\theta = 0.5$ and thus the subjects do not have ESP. A skeptical prior that presumes

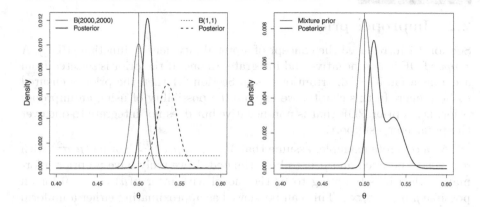

FIGURE 2.5
Prior and posterior distributions for the ESP example. The data are binomial with $Y = 963$ successes in $n = 1800$ trials. The left panel plots the Beta(2000,2000) and Beta(1,1) priors and resulting posteriors, and the right panel plots the mixture prior that equally weights the two beta priors and the resulting posterior.

the subjects do not have ESP should be concentrated around $\theta = 0.5$, say π_1 is the Beta(2000, 2000) PDF. On the other extreme, if we approach the analysis with an open mind then we may select an uninformative prior, say π_2 is the uniform Beta(1, 1) PDF. The mixture prior combines these two extremes,

$$\pi(\theta) = q\pi_1(\theta) + (1 - q)\pi_2(\theta), \tag{2.32}$$

as shown in Figure 2.5 (right) for $q = 0.5$. The mixture prior is simply the average of the two beta prior PDFs in the left panel. The prior has a peak at 0.5 corresponding to π_1, but does not go completely to zero as θ departs from 0.5 because of the addition of the uniform prior π_2. The posterior distribution under the mixture prior (Figure 2.5, right) is also a mixture of two beta distributions. The posterior is a weighted average of the two posterior densities in the left panel. The posterior is bimodal with one mode near 0.51 corresponding to the skeptical prior and a second mode near 0.53 corresponding to the uniform prior. This shape of the density is not possible within the standard beta family.

2.2   Improper priors

Section 1.1 introduced the concept of a probability density function (PDF). A *proper* PDF is non-negative and integrates to one. If the prior is selected from a common family of distributions as in Section 2.1 then the prior is ensured to be proper. In this chapter we explore the possibility of using an improper prior, i.e., a prior PDF that is non-negative but does not integrate to one over the parameter's support.

As a concrete example, assume that $Y_1, ..., Y_n | \mu, \sigma^2 \overset{iid}{\sim} \text{Normal}(\mu, \sigma^2)$ with σ taken to be fixed and μ is to be estimated. If we truly have no prior information then it is tempting to fit the model with prior $\pi(\mu) = c > 0$ for all possible $\mu \in (-\infty, \infty)$. This can be viewed as approximating either a uniform prior with bounds tending to infinity or a Gaussian prior with variance tending to infinity. However, for any positive c, $\int_{-\infty}^{\infty} \pi(\mu) d\mu = \infty$ and so the prior is improper.

Despite the fact that the prior is not a proper PDF, it is still possible to use the density function

$$p(\mu | \mathbf{Y}) \propto f(\mathbf{Y} | \mu) \pi(\mu) \tag{2.33}$$

to summarize uncertainty about μ. That is, we can treat this product of the likelihood and improper prior as a posterior distribution. Applying this rule for the normal mean example gives $\mu | \mathbf{Y} \sim \text{Normal}(\bar{Y}, \sigma^2/n)$, which is a proper PDF as long as $n \geq 1$.

As this example shows, using improper priors can lead to a proper posterior distribution. However, *improper priors must be used with caution*. It must be verified that the resulting posterior distribution is proper, i.e., it must be mathematically proven that the posterior integrates to one over the parameter's support. This step is not needed if the prior is proper because any proper prior leads to a proper posterior. Also, from a conceptual perspective, it is difficult to apply the subjective Bayesian argument that the prior and posterior distributions represent the analyst's uncertainty about the parameters when the prior uncertainty cannot be summarized by a valid PDF. One way to justify treating inference with improper prior as Bayesian inference is to view the posterior under an improper prior as the limit of the posterior for proper priors (with, say, the prior variance increasing to infinity) [78].

An obvious remedy to these concerns is to replace the improper prior with a uniform prior with bounds that almost surely include the true value of a parameter. In fact, a uniform prior was the original prior used by Laplace in his formulation of Bayesian statistics. While this alleviates mathematical problems of ensuring a proper posterior and conceptual problems with Bayesian interpretation, it raises a concern about invariance to parameterization. For example, let $\theta \in (0, 1)$ be the probability of an event and $r = \theta/(1 - \theta) > 0$ be the corresponding odds of the event. A Uniform(0,100) prior for the odds r appears to be uninformative, but in fact the induced prior mean of $\theta = r/(1+r)$ is 0.95, which is clearly not uninformative for θ.

2.3 Objective priors

In an idealized subjective Bayesian analysis the system (physical, biological, economic, etc.) is understood except for the values of a few parameters, and studies are conducted sequentially to refine estimates of the unknown parameters. In this case, the ability to incorporate the current state of knowledge as prior information in the analysis of the current study is a strength of the Bayesian paradigm and should be embraced. However, in many cases there is no prior information, leaving the analyst to select, say, conjugate uninformative priors in an ad hoc manner.

In the absence of prior information, selecting the prior can be viewed as a nuisance to be avoided if possible. Objective priors offer a solution by providing systematic approaches to formulate the prior. Berger [8] argues that objectivity is essential in many analyses, such as obtaining approval from regulatory agencies, where any apparent personal bias can be a huge red flag. Objective methods also provide useful baselines for comparison for subjective methods and make application of Bayesian methods more approachable for non-statisticians [8]. Of course, eliminating the subjectivity of prior selection is only part of the story; there remains subjectivity in selecting the data to be analyzed, the inference to be conducted, the likelihood to be used, and indeed, the form of the objective prior.

In the remainder of this section we discuss several general and objective approaches to setting priors. Jeffreys' prior (Section 2.3.1) is arguably the most common objective Bayes prior and so we describe this method in the most detail. We provide only brief conceptual discussions for the other approaches because deriving the exact expressions of the priors even in moderately complex cases often requires tedious calculus.

2.3.1 Jeffreys' prior

A Jeffreys' prior (JP) is a method to construct a prior distribution that is invariant to reparameterization. This is a necessary condition for a method to be objective. For example, if one statistician places a prior on the standard deviation (σ) and another places a prior on the variance (σ^2) and the two get different results, then the method is subjective because it depends on the choice of parameterization.

The univariate JP for θ is

$$\pi(\theta) \propto \sqrt{I(\theta)} \tag{2.34}$$

where $I(\theta)$ is the expected Fisher information defined as

$$I(\theta) = -\mathrm{E}\left(\frac{d^2 \log f(\mathbf{Y}|\theta)}{d\theta^2}\right) \tag{2.35}$$

and the expectation is with respect to $\mathbf{Y}|\theta$. The key feature of this construction is that it is invariant to transformation. That is, say statistician one uses a JP on θ, denoted $\pi_1(\theta)$, and statistician two places a JP on $\gamma = g(\theta)$, denoted $\pi_2(\gamma)$. Using the change of variables formula, the prior for γ induced by π_1 is

$$\pi_3(\gamma) = \pi_1[h(\gamma)]\left|\frac{dh(\gamma)}{d\gamma}\right| \tag{2.36}$$

where $h(\gamma) = \theta$ is the inverse of g (assuming g is invertable). It can be shown that $\pi_2 = \pi_3$, that is, placing a JP on γ is equivalent to placing a JP on θ and then reparameterizing to γ; this establishes invariance to reparameterization.

Once the likelihood function f has been determined, then the JP is determined; in this sense the prior is objective because any two analysts working with the same likelihood will get the same prior and thus the same posterior. An important point is that the prior does not depend on the data, $\mathbf{Y}$, because the expectation in $I(\theta)$ removes the dependence on the data. Finally, the JP is often improper and therefore before using JP in practice it must be verified that the posterior is a valid PDF that integrates to one (Section 2.2).

JP can also be applied in multivariate problems. The multivariate JP for $\boldsymbol{\theta} = (\theta_1, ..., \theta_p)$ is

$$\pi(\boldsymbol{\theta}) \propto \sqrt{|\mathbf{I}(\boldsymbol{\theta})|} \tag{2.37}$$

where $\mathbf{I}(\boldsymbol{\theta})$ is the $p \times p$ expected Fisher information matrix with (i, j) element

$$-\mathrm{E}\left(\frac{\partial^2 \log f(\mathbf{Y}|\boldsymbol{\theta})}{\partial\theta_i \partial\theta_j}\right). \tag{2.38}$$

As with the univariate JP, this prior is invariant to transformation and can be improper.

Binomial proportion: As a univariate example, consider the binomial model $Y \sim \mathrm{Binomial}(n, \theta)$. The second derivative of the binomial log likelihood is

$$\begin{aligned}
\frac{d^2 \log f(\mathbf{Y}|\theta)}{d\theta^2} &= \frac{d^2}{d\theta^2}\log\binom{n}{Y} + Y\log(\theta) + (n-Y)\log(1-\theta) \\
&= \frac{d}{d\theta}\frac{Y}{\theta} - \frac{n-Y}{1-\theta} \\
&= -\frac{Y}{\theta^2} - \frac{n-Y}{(1-\theta)^2}.
\end{aligned}$$

The information is the expected value of the negative second derivative. Under the binomial model, the expected value of Y is $n\theta$ and since the second derivative is linear in Y the expectation passes into each term as

$$I(\theta) = \frac{\mathrm{E}(Y)}{\theta^2} + \frac{n - \mathrm{E}(Y)}{(1-\theta)^2} = \frac{n}{\theta} + \frac{n}{1-\theta} = \frac{n}{\theta(1-\theta)}. \tag{2.39}$$

The JP is then

$$\pi(\theta) \propto \sqrt{\frac{n}{\theta(1-\theta)}} \propto \theta^{1/2-1}(1-\theta)^{1/2-1}, \tag{2.40}$$

which is the kernel of the Beta(1/2,1/2) PDF. Therefore, the JP for the binomial proportion is $\theta \sim$ Beta(1/2, 1/2). In Section 2.1 we saw that a beta prior was conjugate, and the JP provides an objective way to set the hyperparameters in the beta prior.

Normal model: As a multivariate example, consider the Gaussian model $Y_i|\mu, \sigma^2 \overset{iid}{\sim}$ Normal(μ, σ^2). As shown in Appendix A.3, the bivariate JP is

$$\pi(\mu, \sigma) \propto \frac{1}{(\sigma^2)^{3/2}}. \tag{2.41}$$

This is an improper prior but leads to a proper posterior as long as $n \geq 2$. This prior can be seen as the limit of the product of independent priors $\mu \sim$ Normal$(0, c^2)$ and $\sigma^2 \sim$ InvGamma$(0.5, b)$ as $c \to \infty$ and $b \to 0$. The JP is flat for all values of μ and, similar to the inverse gamma distribution with small shape parameter, the JP peaks at $\sigma^2 = 0$ and drops sharply away from the origin.

Multiple linear regression model: The linear regression model is $\mathbf{Y}|\boldsymbol{\beta}, \sigma^2 \sim$ Normal$(\mathbf{X}\boldsymbol{\beta}, \sigma^2\mathbf{I}_n)$, where the vector of p regression coefficients is $\boldsymbol{\beta} = (\beta_1, ..., \beta_p)$ and the error variance is σ^2. Appendix A.3 shows that the JP is

$$\pi(\boldsymbol{\beta}, \sigma) \propto \frac{1}{(\sigma^2)^{p/2+1}}. \tag{2.42}$$

As in the mean-only model (which is a special case of the regression model), the JP is improper, completely flat in the mean parameters $\boldsymbol{\beta}$, and decays polynomially in the variance with degree that depends on the number of covariates in the model.

2.3.2 Reference priors

Bernardo's reference prior (RP; [10]) is the prior that maximizes the expected difference between the prior and posterior. This is certainly an intuitive definition of an uninformative prior as it ensures that the prior does not overwhelm the data. To act on this conceptual definition requires a measure of the discrepancy between two density functions, and [10] use (ignoring technical problems that require studying sequences of priors) the Kullback–Leibler

(KL) divergence between the prior $\pi(\boldsymbol{\theta})$ and the posterior $p(\boldsymbol{\theta}|\mathbf{Y})$, defined as the expected difference in log densities,

$$KL(p, \pi|\mathbf{Y}) = \mathrm{E}\left\{\log[f(\theta|\mathbf{Y})] - \log[\pi(\theta)]\right\}, \qquad (2.43)$$

where the expectation is with respect to the posterior $p(\boldsymbol{\theta}|\mathbf{Y})$. The KL divergence depends on the data and thus cannot be used to derive a prior. To remove the dependence on the data, the RP is the prior density function π that maximizes the expected information gain $\int KL(p, \pi|\mathbf{Y})m(\mathbf{Y})d\mathbf{Y}$, where $m(\mathbf{Y})$ is the marginal distribution of the data (Table 1.4).

Finding the RP is a daunting optimization problem because the solution is a prior *distribution* and not a scalar or vector. In many cases, the RP differs from the JP. However, it can be shown that if the posterior is approximately normal then the RP is approximately the JP. The two are also equivalent in non-Gaussian cases, for example, if $Y \sim \text{Binomial}(n, \theta)$ the RP is $\theta \sim \text{Beta}(1/2, 1/2)$.

2.3.3   Maximum entropy priors

A maximum entropy (MaxEnt) prior assumes that a few properties of the prior are known and selects the prior with the most entropy in the class of priors with these properties. As an example, say that θ has prior support $\theta \in \{0, 1, 2, ...\}$ and the prior mean is assumed to be $E(\theta) = 1$. There are infinitely many priors for θ that satisfy the constraint that $\mathrm{E}(\theta) = 1$, e.g., $\theta \sim \text{Poisson}(1)$, $\theta \sim \text{Binomial}(2, 0.5)$ or $\theta \sim \text{Binomial}(5, 0.2)$. The MaxEnt prior is the prior that maximizes entropy over the priors with support $\{0, 1, 2, ...\}$ and mean one.

In general, let the constraints be $\mathrm{E}[g_k(\theta)] = c_k$ for functions g_k (mean, variance, exceedance probabilities, etc.) and constants c_k for $k = 1, ..., K$. The entropy for a discrete parameter with prior $\pi(\theta)$ and discrete support $\mathcal{S} = \{\theta_1, \theta_2, ...\}$ is

$$\mathrm{E}\{-\log[\pi(\theta)]\} = -\sum_{i \in \mathcal{S}} \log[\pi(\theta_i)]\pi(\theta_i). \qquad (2.44)$$

The MaxEnt prior is the value of $\{\pi(\theta_1), \pi(\theta_2), ...\}$ that maximizes $-\sum_{i \in \mathcal{S}} \log[\pi(\theta_i)]\pi(\theta_i)$ subject to the K constraints $\sum_{i \in \mathcal{S}} g_k(\theta_i)\pi(\theta_i) = c_k$. MaxEnt priors can be extended to continuous parameters as well.

2.3.4   Empirical Bayes

An empirical Bayesian (EB) analysis uses the data to set the priors. For example, consider the model $Y_i|\theta_i \overset{iid}{\sim} \text{Binomial}(N, \theta_i)$ for $i = 1, ..., n$, where each observation's success probability has prior $\theta_i|a, b \overset{iid}{\sim} \text{Beta}(a, b)$. The goal of this analysis is to estimate the individual probabilities θ_i, and thus there two types of parameters: the parameters of interest $\boldsymbol{\theta} = (\theta_1, ..., \theta_n)$ and the nuisance parameters $\gamma = (a, b)$. Without prior knowledge about γ, a fully Bayesian analysis requires a prior for the nuisance hyperparameters. In contrast, an EB analysis inspects the data to select γ, and then performs a Bayesian analysis of θ as if γ was known all along. For example, in this problem it can be shown (Appendix A.1) that marginally over θ, $Y_i|\gamma \sim$ Beta-Binomial(N, a, b). Therefore, an EB analysis might use the MLE from the beta-binomal model to estimate γ and plug this estimate into the Bayesian analysis for θ.

More formally, consider the model $\mathbf{Y}|\theta \sim f(\mathbf{Y}|\theta)$ and prior $\theta|\gamma \sim \pi(\theta|\gamma)$. The marginal maximum likelihood estimator for the hyperparameter is

$$\hat{\gamma} = \underset{\gamma}{\operatorname{argmax}} \int p(\mathbf{Y}|\theta)\pi(\theta|\gamma)d\theta. \tag{2.45}$$

The EB analysis then conducts a Bayesian analysis of θ with γ set to $\hat{\gamma}$, i.e., $\mathbf{Y}|\theta \sim f(\mathbf{Y}|\theta)$ and $\theta|\gamma = \hat{\gamma} \sim \pi(\theta|\hat{\gamma})$. Although other forms of EB are possible (e.g., a using $\hat{\gamma}$ as the prior mean for γ), this plug-in approach provides a systematic and objective way to set priors.

The disadvantage of EB is obvious: we are using the data twice and ignoring uncertainty about γ. As a result, an EB analysis of θ will often have smaller posterior variances and narrower credible intervals than a fully Bayesian analysis that puts a prior on γ, and thus the statistical properties of the EB analysis can be questioned. Despite this concern, EB remains a useful tool to stabilize computing in high-dimensional problems, especially when it can be argued (e.g., for large data sets) that uncertainty in the nuisance parameters is negligible.

2.3.5 Penalized complexity priors

The penalized complexity (PC) prior [76] is designed to prevent over-fitting caused by using a model that is too complex to be supported by the data. For example, consider the linear regression model

$$\text{Full model}: Y_i|\boldsymbol{\beta}, \sigma^2 \overset{indep}{\sim} \text{Normal}\left(\beta_1 + \sum_{j=2}^{p} X_{ij}\beta_j, \sigma^2\right). \tag{2.46}$$

If the number of predictors p is large then the model becomes very complex. A PC prior might shrink the full model to the base model with $\beta_2 = ... = \beta_p = 0$,

$$\text{Base model}: Y_i|\boldsymbol{\beta}, \sigma^2 \overset{iid}{\sim} \text{Normal}\left(\beta_1, \sigma^2\right). \tag{2.47}$$

Using uninformative priors for the β_j in the full model puts almost no weight on this base model, which could lead to over-fitting if the true model is sparse with many $\beta_j = 0$.

The PC prior is defined using the KL distance between the priors for the full and base models,

$$KL = \int \log\left(\frac{\pi_{full}(\theta)}{\pi_{base}(\theta}\right)\pi_{full}(\theta)d\theta. \tag{2.48}$$

The prior on the distance between models is then $\sqrt{2KL} \sim \text{Gamma}(1, \lambda)$, i.e., an exponential prior with rate λ which peaks at the base model with $KD = 0$ and decreases in complexity KD. Technically this is not an objective Bayesian approach because the user must specify the base model and either λ or a prior for λ. Nonetheless, the PC prior provides a systematic way to set priors for high-dimensional models.

2.4   Exercises

1. Assume $Y_1, ..., Y_n|\mu \overset{iid}{\sim} \text{Normal}(\mu, \sigma^2)$ where σ^2 is fixed and the unknown mean μ has prior $\mu \sim \text{Normal}(0, \sigma^2/m)$.

   (a) Give a 95% posterior interval for μ.

   (b) Select a value of m and argue that for this choice your' 95% posterior credible interval has frequentist coverage 0.95 (that is, if you draw many samples of size n and compute the 95% interval following the formula in (a) for each sample, in the long-run 95% of the intervals will contain the true value of μ).

2. The Major League Baseball player Reggie Jackson is known as "Mr. October" for his outstanding performances in the World Series (which takes place in October). Over his long career he played in 2820 regular-season games and hit 563 home runs in these games (a player can hit 0, 1, 2, ... home runs in a game). He also played in 27 World Series games and hit 10 home runs in these games. Assuming uninformative conjugate priors, summarize the posterior distribution of his home-run rate in the regular season and World Series. Is there sufficient evidence to claim that he performs better in the World Series?

3. Assume that $Y|\theta \sim \text{Exponential}(\theta)$, i.e., $Y|\theta \sim \text{Gamma}(1, \theta)$. Find a conjugate prior distribution for θ and derive the resulting posterior distribution.

4. Assume that $Y|\theta \sim \text{NegBinomial}(\theta, m)$ (see Appendix A.1) and $\theta \sim \text{Beta}(a, b)$.

   (a) Derive the posterior of θ.

   (b) Plot the posterior of θ and give its 95% credible interval assuming $m = 5$, $Y = 10$, and $a = b = 1$.

5. Over the past 50 years California has experienced an average of $\lambda_0 = 75$ large wildfires per year. For the next 10 years you will record the number of large fires in California and then fit a Poisson/gamma model to these data. Let the rate of large fires in this future period, λ, have prior $\lambda \sim \text{Gamma}(a, b)$. Select a and b so that the prior

is uninformative with prior variance around 100 and gives prior probability approximately $\text{Prob}(\lambda > \lambda_0) = 0.5$ so that the prior places equal probability on both hypotheses in the test for a change in the rate.

6. An assembly line relies on accurate measurements from an image-recognition algorithm at the first stage of the process. It is known that the algorithm is unbiased, so assume that measurements follow a normal distribution with mean zero, $Y_i|\sigma^2 \overset{iid}{\sim} \text{Normal}(0, \sigma^2)$. Some errors are permissible, but if σ exceeds the threshold c then the algorithm must be replaced. You make $n = 20$ measurements and observe $\sum_{i=1}^{n} Y_i = -2$ and $\sum_{i=1}^{n} Y_i^2 = 15$ and conduct a Bayesian analysis with $\text{InvGamma}(a, b)$ prior. Compute the posterior probability that $\sigma > c$ for:

   (a) $c = 1$ and $a = b = 0.1$
   (b) $c = 1$ and $a = b = 1.0$
   (c) $c = 2$ and $a = b = 0.1$
   (d) $c = 2$ and $a = b = 1.0$

   For each c, compute the ratio of probabilities for the two priors (i.e. $a = b = 0.1$ and $a = b = 1.0$). Which, if any, of the results are sensitive to the prior?

7. In simple logistic regression, the probability of a success is regressed onto a single covariate X as

$$p(X) = \frac{\exp(\alpha + X\beta)}{1 + \exp(\alpha + X\beta)}. \tag{2.49}$$

   Assuming the covariate is distributed as $X \sim \text{Normal}(0, 1)$ and the parameters have priors $\alpha \sim \text{Normal}(0, c^2)$ and $\beta \sim \text{Normal}(0, c^2)$, find a prior standard deviation c so that the induced prior on the success probability $p(X)$ is roughly $\text{Uniform}(0,1)$. That is, for a given c generate S (say $S = 1,000,000$) samples $(X^{(s)}, \alpha^{(s)}, \beta^{(s)})$ for $s = 1, ..., S$, compute the S success probabilities, and make a histogram of the S probabilities. Repeat this for several c until the histogram is approximately uniform. Report the final value of c and the resulting histogram. Would you call this an uninformative prior for the regression coefficients α and β?

8. The Mayo Clinic conducted a study of $n = 50$ patients followed for one year on a new medication and found that 30 patients experienced no adverse side effects (ASE), 12 experienced one ASE, 6 experienced two ASEs and 2 experienced ten ASEs.

   (a) Derive the posterior of λ for the Poisson/gamma model $Y_1, ..., Y_n|\lambda \overset{iid}{\sim} \text{Poisson}(\lambda)$ and $\lambda \sim \text{Gamma}(a, b)$.

(b) Use the Poisson/gamma model with $a = b = 0.01$ to study the rate of adverse events. Plot the posterior and give the posterior mean and 95% credible interval.

(c) Repeat this analysis with Gamma(0.1,0.1) and Gamma(1,1) priors and discuss sensitivity to the prior.

(d) Plot the data versus the Poisson($\hat{\lambda}$) PMF, where $\hat{\lambda}$ is the posterior mean of λ from part (b). Does the Poisson likelihood fit well?

(e) The current medication is thought to have around one adverse side effect per year. What is the posterior probability that this new medication has a higher side effect rate than the previous mediation? Are the results sensitive to the prior?

9. A smoker is trying to quit and their goal is to abstain for one week because studies have shown that this is the most difficult period. Denote Y as the number of days before the smoker relapses, and θ as the probability that they relapse on any given day of the study.

(a) Argue that a negative binomial distribution (Appendix A.1) for $Y|\theta$ is reasonable. What are the most important assumptions being made and are they valid for this smoking cessation attempt?

(b) Assuming a negative binomial likelihood and Uniform$(0, 1)$ prior for θ, what is the prior probability the smoker will be smoke-free for at least 7 days?

(c) Plot the posterior distribution of θ assuming $Y = 5$.

(d) Given this attempt resulted in $Y = 5$ and that θ is the same in the second attempt, what is the posterior predictive probability the second attempt will result in at least 7 smoke-free days?

10. You are designing a very small experiment to determine the sensitivity of a new security alarm system. You will simulate five robbery attempts and record the number of these attempts that trigger the alarm. Because the dataset will be small you ask two experts for their opinion. One expects the alarm probability to be 0.95 with standard deviation 0.05, the other expects 0.80 with standard deviation 0.20.

(a) Translate these two priors into beta PDFs, plot the two beta PDFs and the corresponding mixture of experts prior with equal weight given to each expert.

(b) Now you conduct the experiment and the alarm is triggered in every simulated robbery. Plot the posterior of the alarm probability under a uniform prior, each experts' prior, and the mixture of experts prior.

11. Let θ be the binomial success probability and $\gamma = \theta/(1 - \theta)$ be the corresponding odds of a success. Use Monte Carlo simulation (you can discard samples with $\gamma > 100$) to explore the effects of parameterization on prior choice in this context.

    (a) If $\theta \sim$ Uniform$(0, 1)$, what is the induced prior on γ?
    (b) If $\theta \sim$ Beta$(0.5, 0.5)$, what is the induced prior on γ?
    (c) If $\gamma \sim$ Uniform$(0, 100)$, what is the induced prior on θ?
    (d) If $\gamma \sim$ Gamma$(1, 1)$, what is the induced prior on θ?
    (e) Would you say any of these priors are simultaneously uninformative for both θ and γ?

12. Say $Y|\mu \sim$ Normal$(\mu, 1)$ and μ has improper prior $\pi(\mu) = 1$ for all μ. Prove that the posterior distribution of μ is a proper PDF.

13. Suppose that the likelihood $f(\mathbf{Y}|\theta) \geq c(\mathbf{Y})$ for all θ. Show that an improper prior on θ will yield an improper posterior.

14. Consider a sample of size n from the mixture of two normal densities given by $f(y|\theta) = 0.5\phi(y) + 0.5\phi(y - \theta)$ where $\phi(z)$ denotes the standard normal density function. Show that any improper prior for θ leads to an improper posterior.

15. Say $Y|\lambda \sim$ Poisson(λ).

    (a) Derive and plot the Jeffreys' prior for λ.
    (b) Is this prior proper?
    (c) Derive the posterior and give conditions on Y to ensure it is proper.

16. Say $Y|\lambda \sim$ Gamma$(1, \lambda)$.

    (a) Derive and plot the Jeffreys' prior for λ.
    (b) Is this prior proper?
    (c) Derive the posterior and give conditions on Y to ensure it is proper.

17. Assume the model $Y|\lambda \sim$ Poisson(λ) and we observe $Y = 10$. Conduct an objective Bayesian analysis of these data.

18. The data in the table below are the result of a survey of commuters in 10 counties likely to be affected by a proposed addition of a high occupancy vehicle (HOV) lane.

County	Approve	Disapprove	County	Approve	Disapprove
1	12	50	6	15	8
2	90	150	7	67	56
3	80	63	8	22	19
4	5	10	9	56	63
5	63	63	10	33	19

(a) Analyze the data in each county separately using the Jeffreys' prior distribution and report the posterior 95% credible set for each county.

(b) Let $\hat{p}_i$ be the sample proportion of commuters in county i that approve of the HOV lane (e.g., $\hat{p}_1 = 12/(12 + 50) = 0.194$). Select a and b so that the mean and variance of the Beta(a, b) distribution match the mean and variance of the sample proportions $\hat{p}_1, ..., \hat{p}_{10}$.

(c) Conduct an empirical Bayesian analysis by computing the 95% posterior credible sets that results from analyzing each county separately using the Beta(a, b) prior you computed in (b).

(d) How do the results from (a) and (c) differ? What are the advantages and disadvantages of these two analyses?

3

Computational approaches

CONTENTS

Computing is a key ingredient to any modern statistical application and a Bayesian analysis is no different. As of the late 1980's, the application of Bayesian methods was limited to small problems where conjugacy (Section 2.1) led to a simple posterior or the number of parameters was low enough to permit numerical integration (e.g., Section 3.1.2). However, with the advent of Markov Chain Monte Carlo (MCMC) methods (which were developed in the 1950s through 1970s) the popularity of Bayesian methods exploded as it became possible to fit realistic models to complex data sets. This chapter is dedicated to the core Bayesian computational tools that led to this resurgence.

Bayesian methods are often associated with heavy computing. This is not an inherent property of the Bayesian paradigm. Bayesian computing is designed to estimate the posterior distribution, which is analogous to frequentist computing estimating the sampling distribution. Both problems are challenging, but frequentists often make use of large-sample normal approximations to simplify the problem. There is nothing to prohibit a Bayesian from making similar approximations, and this is the topic of Section 3.1.3. However, these approximations are not common because MCMC produces much richer output, and the accuracy of the approximation is not limited by the sample size of the data, but rather the MCMC approximation can be made arbitrarily precise even for non-Gaussian posteriors by increasing the computational effort.

Bayesian computation boils down to summarizing a high-dimensional posterior distribution (Section 1.4). In this chapter we cover methods that span the spectrum from summaries that are fast to compute (e.g., MAP estimation) but produce only a point estimate to slower methods (e.g., MCMC) that permit full uncertainty quantification. Just as a user of maximum likelihood analysis need not be an expert on the theory and practice of iteratively reweighted least squares optimization, there is software (Section 3.3) available to implement most Bayesian analyses without the user possessing a deep understanding of the algorithms being used under the hood. However, at a minimum, the user should have sufficient understanding of the underlying algorithms to properly inspect the output to determine if it is sensible. In Sections 3.1 and 3.2, we outline the fundamental algorithms used in Bayesian computing and work a few small examples for illustration. However, for the remainder of the book, we will use the JAGS package introduced in Section 3.3 for computation to avoid writing MCMC code.

3.1 Deterministic methods

3.1.1 Maximum a posteriori estimation

Most of the computational methods discussed in this chapter attempt to summarize the entire posterior distribution. This allows for full quantification of uncertainty about the parameters and leads to inferences such as the posterior probability that a hypothesis is true. However, summarizing the full posterior distribution can be difficult and in some settings unnecessary. For example, in machine-learning, statistical estimation is used like a filtering process where the sole purpose is the prediction of future values using estimates from the current data, and accounting for uncertainty and accessing statistical significance are not high priorities. In this case, a single point estimate may be a sufficient summary of the posterior.

The MAP (Section 1.4.1) point estimator is

$$\hat{\boldsymbol{\theta}}_{MAP} = \arg\max_{\boldsymbol{\theta}} \log[p(\boldsymbol{\theta}|\mathbf{Y})] = \arg\max_{\boldsymbol{\theta}} \log[f(\mathbf{Y}|\boldsymbol{\theta})] + \log[\pi(\boldsymbol{\theta})]. \qquad (3.1)$$

The MAP estimator requires optimization of the posterior, unlike the posterior mean which requires integration. Typically optimization is faster than integration, especially in high-dimensions, and therefore MAP estimation can be applied even for hard problems.

The MAP solution can be found using calculus or numerical optimization [62]. The simplest optimization algorithm is gradient ascent (or gradient descent to minimize a function), which is an iterative algorithm that begins with an initial value $\theta^{(0)}$ and updates the parameters at iteration t using the rule

$$\theta^{(t)} = \theta^{(t-1)} + \gamma \nabla \left(\theta^{(t-1)} \right),$$

where the step size γ is a tuning parameter, the gradient vector of the log posterior is $\nabla(\theta) = [\nabla_1(\theta), ..., \nabla_p(\theta)]^T$ for

$$\nabla_j(\theta) = \frac{\partial}{\partial \theta_j} \{\log[f(\mathbf{Y}|\theta)] + \log[\pi(\theta)]\}.$$

This step is repeated until convergence, i.e., $\theta^{(t)} \approx \theta^{(t-1)}$. Gradient ascent is most effective when the posterior is convex, and typically requires several starting values to ensure convergence.

There is a rich literature on optimization in statistical computing. Some other useful methods include Newton's method, the expectation-maximization (EM) algorithm, the majorize-minimization (MM) algorithm, genetic algorithms, etc. R has many optimization routines including the general-purpose algorithm `optim`, as illustrated in Listing 3.1 and Figure 3.1 for the model $Y_i|\mu,\sigma \overset{iid}{\sim} \text{Normal}(\mu, \sigma^2)$ with priors $\mu \sim \text{Normal}(0, 100^2)$ and $\sigma \sim \text{Unif}(0, 10)$ and data $\mathbf{Y} = (2.68, 1.18, -0.97, -0.98, -1.03)$.

3.1.2 Numerical integration

Many posterior summaries of interest can be written as p-variate integrals over the posterior, including the posterior means, covariances and probabilities of hypotheses. For example, for θ_1, the marginal posterior mean, variance and probability that θ_1 exceeds constant c are

$$\text{E}(\theta_1|\mathbf{Y}) = \int \theta_1 p(\theta|\mathbf{Y}) d\theta \tag{3.2}$$

$$\text{Var}(\theta_1|\mathbf{Y}) = \int [\theta_1 - \text{E}(\theta_1|\mathbf{Y})]^2 p(\theta|\mathbf{Y}) d\theta$$

$$\text{Prob}(\theta_1 > c|\mathbf{Y}) = \int_c^\infty \int \cdots \int p(\theta|\mathbf{Y}) d\theta_1 d\theta_2, ..., d\theta_p.$$

If these summaries are sufficient to describe the posterior, then numerical integration can be used for Bayesian computing.

All of the summaries in (3.2) can be written as $\text{E}[g(\theta)]$ for some function g, e.g., $\text{Prob}(\theta_1 > c|\mathbf{Y})$ uses $g(\theta) = I(\theta_1 > c)$ where $I(\theta_1 > c) = 1$ if $\theta_1 > c$ and zero otherwise. Assume that a grid of points $\theta_1^*, ..., \theta_m^*$ covers the range of θ with non-zero posterior density. Then

$$\text{E}[g(\theta)] = \int g(\theta) p(\theta|Y) d\theta \approx \sum_{j=1}^m g(\theta_j) W_j \tag{3.3}$$

where W_j is the weight given to grid point j. To approximate the posterior mean of $g(\theta)$ the weights should be related to the posterior PDF

Listing 3.1
Numerical optimization and integration to summarize the posterior.

```
1   library(cubature)
2   Y <- c(2.68, 1.18, -0.97, -0.98, -1.03) # Data
3
4   # Evaluate the density on the grid for plotting
5   m     <- 50
6   mu    <- seq(-4,6,length=m)
7   sigma <- seq(0,10,length=m)
8   theta <- as.matrix(expand.grid(mu,sigma))
9   D     <- dnorm(theta[,1],0,100)*dunif(theta[,2],0,10) # Prior
10  for(i in 1:length(Y)){ # Likelihood
11    D <- D * dnorm(Y[i],theta[,1],theta[,2])
12  }
13  W     <- matrix(D/sum(D),m,m)
14
15  # MAP estimation
16  neg_log_post <- function(theta,Y){
17    log_like  <- sum(dnorm(Y,theta[1],theta[2],log=TRUE))
18    log_prior <- dnorm(theta[1],0,100,log=TRUE)+
19                 dunif(theta[2],0,10,log=TRUE)
20    return(-log_like-log_prior)}
21
22  inits <- c(mean(Y),sd(Y))
23  MAP   <- optim(inits,neg_log_post,Y=Y,
24              method = "L-BFGS-B", # Since the prior is bounded
25              lower = c(-Inf,0), upper = c(Inf,10))$par
26
27  # Compute the posterior mean
28  post <- function(theta,Y){
29    like  <- prod(dnorm(Y,theta[1],theta[2]))
30    prior <- dnorm(theta[1],0,100)*dunif(theta[2],0,10)
31    return(like*prior)}
32
33  g0 <- function(theta,Y){post(theta,Y)}
34  g1 <- function(theta,Y){theta[1]*post(theta,Y)}
35  g2 <- function(theta,Y){theta[2]*post(theta,Y)}
36  m0 <- adaptIntegrate(g0,c(-5,0.01),c(5,5),Y=Y)$int #constant m(Y)
37  m1 <- adaptIntegrate(g1,c(-5,0.01),c(5,5),Y=Y)$int
38  m2 <- adaptIntegrate(g2,c(-5,0.01),c(5,5),Y=Y)$int
39  pm <- c(m1,m2)/m0
40
41  # Make the plot
42  image(mu,sigma,W,col=gray.colors(10,1,0),
43        xlab=expression(mu),ylab=expression(sigma))
44  points(theta,cex=0.1,pch=19)
45  points(pm[1],pm[2],pch=19,cex=1.5)
46  points(MAP[1],MAP[2],col="white",cex=1.5,pch=19)
47  box()
```

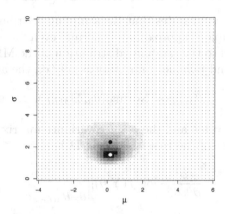

FIGURE 3.1
Numerical integration and optimization. The small points are the grid
points $\boldsymbol{\theta}_j = (\mu_j^*, \sigma_j^*)$ and the color of the shading is the posterior density for
the model $Y_i | \mu, \sigma \sim \text{Normal}(\mu, \sigma^2)$ with $\mathbf{Y} = (2.68, 1.18, -0.97, -0.98, -1.03)$
and priors $\mu \sim \text{Normal}(0, 100^2)$ and $\sigma \sim \text{Uniform}(0, 10)$. The large black
dot is the approximate posterior mean found using numerical integration and
the large white dot is the approximate posterior mode found using numerical
optimization.

$D_j = f(\mathbf{Y}|\boldsymbol{\theta}_j^*)\pi(\boldsymbol{\theta}_j^*)$. The normalizing constant $m(\mathbf{Y})$ cannot be computed and so the posterior is normalized at the grid points as $W_j = D_j/\sum_{l=1}^m D_l$.

This provides a very simple approximation to the posterior of $\boldsymbol{\theta}$. Of course, more precise numerical integration can be used, including those in the R function `adaptIntegrate` as illustrated in Listing 3.1 and Figure 3.1. However, we do not focus on these methods because they do not scale well with the number of parameters, p. For example, if we use a grid of 20 points for each of the $p = 10$ parameters then there are $m = 20^{10}$ points in the expanded grid, which is over a trillion points!

3.1.3   Bayesian central limit theorem (CLT)

The central limit theorem states that the sampling distribution of the sample mean for a large sample is approximately normal even for non-Gaussian data. This can be extended to the sampling distribution of the MLE. Many frequentist standard error and p-value computations rely on the approximation

$$\hat{\boldsymbol{\theta}}_{MLE} \sim \text{Normal}(\boldsymbol{\theta}_0, \hat{\boldsymbol{\Sigma}}_{MLE}), \tag{3.4}$$

where $\boldsymbol{\theta}_0$ is the true value and the $p \times p$ covariance matrix $\hat{\boldsymbol{\Sigma}}_{MLE} = (-\mathbf{H})^{-1}$ and the (j, k) element of the Hessian matrix $\mathbf{H}$ is

$$\left.\frac{\partial^2}{\partial\theta_j\partial\theta_k} \log[f(\mathbf{Y}|\boldsymbol{\theta})]\right|_{\boldsymbol{\theta}=\hat{\boldsymbol{\theta}}_{MLE}}. \tag{3.5}$$

To define the analogous Bayesian approximation, suppose that $Y_i|\boldsymbol{\theta} \overset{indep}{\sim} f(Y_i|\boldsymbol{\theta})$ for $i = 1, ..., n$ and $\boldsymbol{\theta} \sim \pi(\boldsymbol{\theta})$. Under general conditions (see Section 7.2.2), for large n the posterior is approximately

$$\boldsymbol{\theta}|\mathbf{Y} \sim \text{Normal}\left(\hat{\boldsymbol{\theta}}_{MAP}, \hat{\boldsymbol{\Sigma}}_{MAP}\right) \tag{3.6}$$

where $\hat{\boldsymbol{\Sigma}}_{MAP} = (-\mathbf{H})^{-1}$ is defined the same way as $\hat{\boldsymbol{\Sigma}}_{MLE}$ except that the Hessian $\mathbf{H}$ is

$$\left.\frac{\partial^2}{\partial\theta_j\partial\theta_k} \log[p(\boldsymbol{\theta}|\mathbf{Y})]\right|_{\boldsymbol{\theta}=\hat{\boldsymbol{\theta}}_{MAP}}, \tag{3.7}$$

which will be similar to $\hat{\boldsymbol{\Sigma}}_{MLE}$ for large samples. Of course, this is not appropriate if the parameters are discrete, but the Bayesian CLT can still be used in some cases where the observations are dependent.

For example, consider the beta-binomial model $Y|\theta \sim \text{Binomial}(n, \theta)$ with Jeffreys prior $\theta \sim \text{Beta}(1/2, 1/2)$. The MAP estimate is $\hat{\theta}_{MAP} = A/(A + B)$ and the approximate posterior variance is

$$\left[\frac{A}{\hat{\theta}_{MAP}^2} + \frac{B}{(1 - \hat{\theta}_{MAP})^2}\right]^{-1},$$

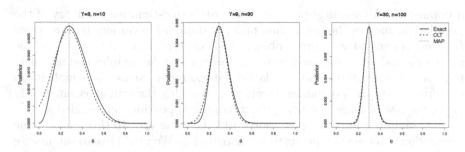

FIGURE 3.2
Illustration of the Bayesian CLT. The exact Beta($Y + 1/2, n + 1/2$) posterior versus the Gaussian approximation for the model $Y|\theta \sim$ Binomial(n, θ) with prior $\theta \sim$ Beta($1/2, 1/2$) for various Y and n.

where $A = Y - 0.5$ and $B = n - Y - 0.5$. Of course, there is no need for an approximation in this simple case because the exact posterior is $\theta|Y \sim$ Beta($Y + 1/2, n + 1/2$), but Figure 3.2 illustrates that the Gaussian approximation works well for large n.

As discussed in Section 3.1.1, if the prior is uninformative then the MAP and MLE estimates are similar, and these approximations suggest that the entire posterior will be similar. This is the first glimpse at a recurring theme: Bayesian and maximum likelihood methods give similar results for large sample sizes irrespective of the prior.

Advantages of the Bayesian CLT approximation are that it replaces integration (Section 3.1.2) with differentiation, and is thus easier to compute, especially for large p, and the method is deterministic rather than stochastic. A disadvantage is that approximation may not be accurate for small or even moderate samples sizes. However, more elaborate approximation methods have been proposed that combine numerical integration with Gaussian approximations, such as the integrated nested Laplace approximation (INLA) of [73] (Appendix A.4). Alternatively, in variational Bayesian inference [46] the user postulates a parametric (potentially non-Gaussian) posterior distribution and then solves for the parameters in the postulated posterior that give the best approximation to the true posterior.

3.2 Markov chain Monte Carlo (MCMC) methods

Monte Carlo (MC) methods are appealing to statisticians because they mirror the fundamental statistical concept of using a sample to make inferences about a population. A canonical problem in statistics to estimate a summary of the population, such as the population mean or standard deviation. In most cases (i.e., not a census) we cannot observe the entire population to compute the mean or standard deviation directly, so instead we take samples and use them to make inference about the population. For example, we use the sample mean to approximate the population mean and, assuming the sample is sufficiently large, the law of large numbers guarantees the approximation is reliable.

MC sampling from the posterior works the same way. In this analogy, the population of interest is the posterior distribution. We would like to summarize the posterior using its mean and variance, but in most cases these posterior summaries cannot be computed directly, and so we take a sample from the posterior and use the MC sample mean to approximate the posterior mean and the MC sample variance to approximate the posterior variance. Assuming the MC sample is sufficiently large, then the approximation is reliable. This holds for virtually any posterior distribution and any summary of the posterior, even those that are high-dimensional integrals.

Assume that we have generated S posterior samples, $\boldsymbol{\theta}^{(1)}, ..., \boldsymbol{\theta}^{(S)} \sim p(\boldsymbol{\theta}|\mathbf{Y})$, where sample s is $\boldsymbol{\theta}^{(s)} = (\theta_1^{(s)}, ..., \theta_p^{(s)})$. These draws can be used to approximate any posterior summary discussed in Section 1.4. For a univariate model with $p = 1$, we could approximate the posterior density with a histogram of the S samples, the posterior mean with the MC sample mean $E(\theta_1|\mathbf{Y}) \approx \sum_{s=1}^{S} \theta_1^{(s)}/S$, the posterior variance with the MC sample variance of the S samples, the probability that $\theta_1 > 0$ with the MC sample proportion with $\theta_1^{(s)} > 0$, etc.

To illustrate convergence of MC sampling, consider the Poisson-gamma model $Y|\theta \sim \text{Poisson}(N\theta)$ with prior $\theta \sim \text{Gamma}(a, b)$. The posterior is then $\theta|Y \sim \text{Gamma}(Y + a, N + b)$. Figure 3.3 assumes $N = 10$, $Y = 8$ and $a = b = 0.1$ and plots the approximate posterior mean and probability that $\theta > 0.5$ as a function of the number of MC iterations. The sample mean and proportion are noisy for a small number of MC samples, but as the number of samples increases they converge to the true posterior mean and probability that $\theta > 0.5$.

For multivariate models with $p > 1$ parameters, the samples $\theta_j^{(1)}, ..., \theta_j^{(S)}$ follow the marginal posterior distribution of θ_j, $p(\theta_j|\mathbf{Y})$. Critically, we do not need to analytically integrate $p(\theta_j|\mathbf{Y}) = \int f(\boldsymbol{\theta}|\mathbf{Y})d\theta_1...d\theta_{j-1}d\theta_{j+1}...d\theta_p$. Because each sample consists of a random draw from all parameters, MC sampling automatically produces samples from the marginal distribution θ_j accounting for uncertainty in the other parameters.

In addition to the marginal distribution, MC sampling can be used to approximate the posterior of parameters defined as transformations of the original parameters. For example, we can approximate the posterior of $\gamma = g(\theta_1)$ for a function g using the MC samples $\gamma^{(1)}, ..., \gamma^{(S)}$ where $\gamma^{(s)} = g(\theta_1^{(s)})$, that is, we simply transform each sample and these transformed samples approximate the posterior distribution of γ. The posterior of the transformed param-

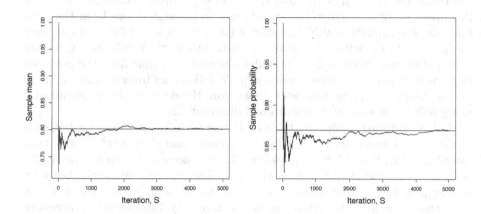

FIGURE 3.3
Convergence of a Monte Carlo approximation. Assuming $\theta^{(s)} \overset{iid}{\sim}$ Gamma(8.1, 10.1), the plot below gives the sample mean $\sum_{s=1}^{S} \theta^{(s)}/S$ and sample proportion $\sum_{s=1}^{S} I(\theta^{(s)} > 0.5)/S$ by the number of samples, S; the horizontal lines are the true values.

eter is summarized like any other, e.g., $\mathrm{E}(\gamma|\mathbf{Y}) \approx \sum_{s=1}^{S} \gamma^{(s)}/S$. As another example, say we want to test if $\Delta = \theta_1 - \theta_2 > 0$. The posterior of Δ is approximated by the S samples $\Delta^{(s)} = \theta_1^{(s)} - \theta_2^{(s)}$, and thus $\mathrm{Prob}(\Delta > 0|\mathbf{Y})$ is approximated using the sample proportion of the S samples for which $\Delta^{(s)} > 0$.

Once we have posterior samples, summarizing the posterior or even complicated functions of the posterior is straightforward and this one of the appeals of MC sampling. However, generating valid samples from the posterior distribution is not always straightforward. We will focus on two sampling algorithms: Gibbs sampling and Metropolis–Hastings sampling. Gibbs sampling is preferred when it is possible to directly sample from the conditional distribution of each parameter. Metropolis–Hastings is a generalization to more complicated problems. We also briefly mention other more advanced algorithms in Appendix A.4 including delayed rejection and adaptive Metropolis (DRAM), the Metropolis-adjusted Langevin algorithm, Hamiltonian MC (HMC) and slice sampling.

3.2.1 Gibbs sampling

As a motivating example, consider the Gaussian model

$$Y_1, ..., Y_n \overset{iid}{\sim} \mathrm{Normal}(\mu, \sigma^2) \tag{3.8}$$

where the priors are $\mu \sim \text{Normal}(\gamma, \tau^2)$ and $\sigma^2 \sim \text{InvGamma}(a, b)$. To study the posterior of $\theta = (\mu, \sigma^2)$ we would like to make S draws from the joint posterior distribution $p(\theta|\mathbf{Y})$. Section 2.1.3 provides a means to sample from the posterior of μ with σ^2 assumed known since $\mu|\sigma^2, \mathbf{Y}$ follows a Gaussian distribution, and Section 2.1.4 provides a means to sample from the posterior of σ^2 with μ assumed known since $\sigma^2|\mu, \mathbf{Y}$ follows an inverse gamma distribution. Gibbs sampling generates draws from the desired joint posterior of θ using only these univariate conditional distributions.

Gibbs sampling (proposed by [34]) begins with initial values for both parameters and proceeds by alternating between sampling $\mu|\sigma^2, \mathbf{Y}$ and then sampling $\sigma^2|\mu, \mathbf{Y}$ until S samples have been collected (for an extensive discussion about selecting S, see Section 3.4). For example, we might begin by setting μ to the sample mean of $\mathbf{Y}$ and σ^2 to the sample variance of $\mathbf{Y}$ (for an extensive study of initialization, see Section 3.4). Implementing successive steps requires deriving the full conditional posterior distributions, i.e., the distribution of one parameter conditioned on the data and all other parameters. Following Section 2.1.3 and 2.1.4 (with slight modification to the prior distributions), the full conditional posterior distributions are

$$\mu|\sigma^2, \mathbf{Y} \quad \sim \quad \text{Normal}\left(\frac{\sum_i^n Y_i/\sigma^2 + \gamma/\tau^2}{n/\sigma^2 + 1\tau^2}, \frac{1}{n/\sigma^2 + 1/\tau^2}\right)$$

$$\sigma^2|\mu, \mathbf{Y} \quad \sim \quad \text{InvGamma}\left(n/2 + a, \sum_{i=1}^n (Y_i - \mu)^2/2 + b\right).$$

This results in S posterior samples $\theta^{(1)}, ..., \theta^{(S)}$ to summarize the posterior, where $\theta^{(s)} = (\mu^{(s)}, \sigma^{2(s)})$. Listing 3.2 provides R code to perform these steps using the same data as in Section 1.1.2. Figure 3.4 (bottom row) plots the draws from the joint posterior and marginal density for μ, which both closely resemble Figure 1.11.

Algorithm 1 provides a general recipe for Gibbs sampling. The algorithm begins with an initial value for the parameter vector and each iteration consists of a sequence of updates from each parameter's full conditional distribution. Because each step cycles through the parameters and updates them given the current values of all other parameters, the samples are not independent. All sampling is performed conditionally on the previous iteration's value, and so the samples form a specific type of stochastic process called a Markov chain, hence the name Markov chain Monte Carlo (MCMC) sampling. There are MC samplers that attempt to sample independent draws from the posterior (e.g., rejection sampling and approximate Bayesian computing (ABC) [53]), but MCMC is preferred for high-dimensional problems.

The beauty of this algorithm is that it reduces the difficult problem of sampling from a multivariate distribution down to a sequence of simper univariate problems. This assumes that the full conditional distributions are easy to sample from, but as we will see in the following examples, even for high-

Listing 3.2
Gibbs sampling for the Gaussian model with unknown mean and variance.

```
1   # Load the data
2
3   Y <- c(2.68,1.18,-0.97,-0.98,-1.03)
4   n <- length(Y)
5
6   # Create an empty matrix for the MCMC samples
7
8   S                   <- 25000
9   samples             <- matrix(NA,S,2)
10  colnames(samples) <- c("mu","sigma")
11
12  # Initial values
13
14  mu   <- mean(Y)
15  sig2 <- var(Y)
16
17  # priors: mu ~ N(gamma,tau), sig2 ~ InvG(a,b)
18
19  gamma <- 0
20  tau   <- 100^2
21  a     <- 0.1
22  b     <- 0.1
23
24  # Gibbs sampling
25
26  for(s in 1:S){
27    P     <- n/sig2 + 1/tau
28    M     <- sum(Y)/sig2 + gamma/tau
29    mu    <- rnorm(1,M/P,1/sqrt(P))
30
31    A     <- n/2 + a
32    B     <- sum((Y-mu)^2)/2 + b
33    sig2 <- 1/rgamma(1,A,B)
34
35    samples[s,]<-c(mu,sqrt(sig2))
36  }
37
38  # Plot the joint posterior and marginal of mu
39  plot(samples,xlab=expression(mu),ylab=expression(sigma))
40  hist(samples[,1],xlab=expression(mu))
41
42  # Posterior mean, median and credible intervals
43  apply(samples,2,mean)
44  apply(samples,2,quantile,c(0.025,0.500,0.975))
```

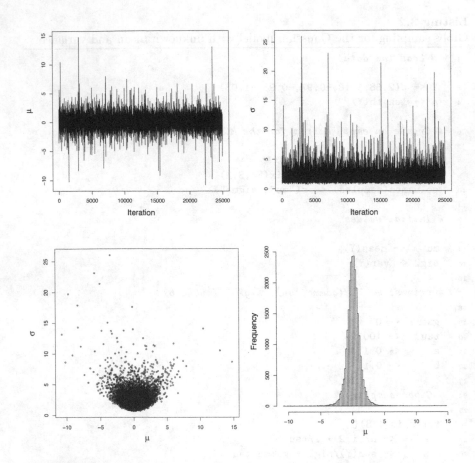

FIGURE 3.4
Summary of posterior samples from MCMC for the Gaussian model with unknown mean and variance. The first row gives trace plots of the samples for μ and σ, the second row shows the samples from the joint posterior of (μ, σ) and the marginal posterior of μ.

Algorithm 1 Gibbs Sampling

1: Initialize $\boldsymbol{\theta}^{(0)} = (\theta_1^{(0)}, ..., \theta_p^{(0)})$
2: **for** $s = 1, ..., S$ **do**
3:     **for** $j = 1, ..., p$ **do**
4:         sample $\theta_j^{(s)} \sim p_j(\theta_j | \theta_1^{(s)}, ..., \theta_{j-1}^{(s)}, \theta_{j+1}^{(s-1)}, ..., \theta_p^{(s-1)}, \mathbf{Y})$
5:     **end for**
6: **end for**

---

dimensional problems with large p, the full conditional distributions often follow familiar conjugacy pairs (Section 2.1) which are conducive to sampling.

Before continuing with other examples, we pause and discuss the theoretical properties of this algorithm. Why does it work? That is, why does repeated sampling from full conditional distributions lead to samples from the joint posterior distribution? Surely we cannot trust that the initial value is a sample from the posterior because it is selected subjectively by the user. It would also be dangerous to trust that the first random sample follows the posterior distribution because it depends on the initial values, which might be far from the posterior. Appendix A.3 provides an argument using stochastic process theory that (1) under general conditions, the samples produced by the algorithm converge to the posterior distribution for any choice of initial values and (2) once a sample is drawn from the posterior, all subsequent samples are also samples from the posterior.

The theoretical convergence arguments dictate how Gibbs sampling is used in practice. The user follows the chain until it has converged, and discards all previous samples from this burn-in period. Convergence is often assessed by visual inspection of the trace plots (i.e., the a plot of the samples by iteration number), but formal measures are available (Section 3.4). All remaining samples are used to summarize the posterior. It is important to remember that unlike other optimization algorithms, we do not expect Gibbs sampling to converge to a single optimal value. Rather, we hope that after burn-in the algorithm produces samples from the posterior distribution. That is, we hope the samples generated by the algorithm converge in distribution to the posterior, and do not get stuck in one place. Figure 3.5 provides idealized output where convergence clearly occurs around iteration 1000 for both parameters and the remaining samples consistently draw from the same distribution after this point and thus the trace plot resembles a "bar code" or "caterpillar."

Bivariate normal example: We will use Gibbs sampling to make draws from posterior distributions, but in fact it can be used to sample from any distribution (so long as we can sample from the full conditional distributions). For example, say $\theta = (U, V)$ is bivariate normal with mean zero and variance 1 for both parameters and correlation ρ between the parameters. This is not a posterior distribution because we are not conditioning on data, but in this toy example we are using Gibbs sampling to draw from this bivariate normal distribution to get a feel for the algorithm and how it converges to the target distribution. As given in (1.26) (see Figure 1.7), the full conditional distributions are $U|V \sim \text{Normal}(\rho V, 1 - \rho^2)$ and $V|U \sim \text{Normal}(\rho U, 1 - \rho^2)$. Given initial values $(U^{(0)}, V^{(0)})$, the first three iterations are:

1a   Draw $U^{(1)}|V^{(0)} \sim \text{Normal}(\rho V^{(0)}, 1 - \rho^2)$

1b   Draw $V^{(1)}|U^{(1)} \sim \text{Normal}(\rho U^{(1)}, 1 - \rho^2)$

2a   Draw $U^{(2)}|V^{(1)} \sim \text{Normal}(\rho V^{(1)}, 1 - \rho^2)$

2b   Draw $V^{(2)}|U^{(2)} \sim \text{Normal}(\rho U^{(2)}, 1 - \rho^2)$

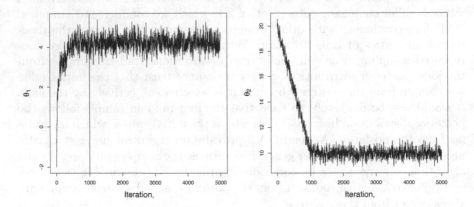

FIGURE 3.5
Gibbs sampling trace plots. These trace plots of the iteration number s by the posterior samples $\theta_j^{(s)}$ for $j = 1, 2$ show convergence at iteration 1000, which is denoted with a vertical line.

    3a Draw $U^{(3)}|V^{(2)} \sim \text{Normal}(\rho V^{(2)}, 1 - \rho^2)$

    3b Draw $V^{(3)}|U^{(3)} \sim \text{Normal}(\rho U^{(3)}, 1 - \rho^2)$.

The algorithm continues until S draws have been made.

    It can be shown that the distribution after s iterations is

$$U^{(s)} \sim \text{Normal}(\rho^{2s-1}V^{(0)}, 1 - \rho^{4s-2}).$$

The true marginal distribution of U is Normal$(0, 1)$, and since $\rho^{2s-1} \approx 0$ and $\rho^{4s-2} \approx 0$ for large s, for any initial value the MCMC posterior is very close but never exactly equal to the true posterior. Convergence is immediate if $\rho = 0$ and slow if $|\rho| \approx 1$, illustrating how cross-correlation between parameters can hinder MCMC convergence (e.g., Figure 3.8 and the discussion of blocked Gibbs sampling below). Nonetheless, for any ρ and large S the approximation is surely sufficient and can be made arbitrarily precise by increasing S.

    Constructing a Gibbs sampler for NFL concussions data: In Section 2.1 we analyzed the NFL concussion data. The data consist of the number of concussions in the years 2012–2015, $Y_1 = 171$, $Y_2 = 152$, $Y_3 = 123$, and $Y_4 = 199$, respectively. In Section 2.1 we analyzed the years independently, but here we analyze all years simultaneously with the model

$$Y_i|\lambda_i \overset{indep}{\sim} \text{Poisson}(N\lambda_i) \quad \lambda_i|\gamma \overset{indep}{\sim} \text{Gamma}(1, \gamma) \quad \gamma \sim \text{Gamma}(a, b) \quad (3.9)$$

where $N = 256$ is the number of games per season. The model has five unknown parameters $\boldsymbol{\theta} = (\lambda_1, ..., \lambda_4, \gamma)$ and posterior

$$p(\lambda_1, ..., \lambda_4, \gamma | \mathbf{Y}) \propto \left[\prod_{i=1}^{4} f(Y_i | \lambda_i) \pi(\lambda_i | \gamma) \right] \pi(\gamma), \qquad (3.10)$$

where f is the Poisson PMF and π is the gamma PDF.

The Gibbs sampler requires the full conditional posterior for each of the five parameters. We first compute the full conditional of λ_1 given all the other parameters. For this computation, we only need to consider terms that depend on λ_1, and all other terms can be absorbed in the normalizing constant. For λ_1 this gives

$$p(\lambda_1, | \lambda_2, \lambda_3, \lambda_4, \gamma, \mathbf{Y}) \propto f(Y_1 | \lambda_1) \pi(\lambda_1 | \gamma). \qquad (3.11)$$

This is exactly the form of the posterior for the Poisson-gamma conjugacy pair studied in Section 2.1, and so the full conditional of λ_1 is Gamma$(Y_1 + 1, N + \gamma)$; the full conditionals for the other λ_j are similar.

The terms in the likelihood $f(Y_i | \lambda_i)$ do not depend on γ, and so the full conditional is

$$p(\gamma | \lambda_1, \lambda_2, \lambda_3, \lambda_4, \mathbf{Y}) \propto \left[\prod_{i=1}^{4} \pi(\lambda_i | \gamma) \right] \pi(\gamma). \qquad (3.12)$$

When viewed as a function of γ, the Gamma$(1, \gamma)$ prior $\pi(\lambda_i | \gamma) = \gamma \lambda_i \exp(-\gamma \lambda_i) \propto \gamma \exp(-\gamma \lambda_i)$. Therefore, the full conditional reduces to

$$p(\gamma | \lambda_1, \lambda_2, \lambda_3, \lambda_4, \mathbf{Y}) \propto \left[\prod_{i=1}^{4} \gamma \exp(-\gamma \lambda_i) \right] \gamma^{a-1} \exp(-\gamma b)$$

$$\propto \gamma^{4+a-1} \exp \left[-\gamma \left(\sum_{i=1}^{4} \lambda_i + b \right) \right],$$

and thus the update for γ is Gamma$(4 + a, \sum_{i=1}^{4} \lambda_i + b)$. Note that the update does not depend on the data, $\mathbf{Y}$. However, the data are informative about γ indirectly because they largely determine the posterior of the λ_j which in turn influence the posterior of γ.

Listing 3.3 provides the code that generates the results in Figure 3.6. Convergence for γ (left panel of Figure 3.6) is immediate (as it is for the λ_i, not shown), and so we do not need a burn-in period and all samples are used to summarize the posterior. The posterior distribution from this analysis of all years (right panel of Figure 3.6) is actually quite similar to the one-year-at-a-time analysis in Figure 2.2.

Constructing a blocked Gibbs sampler for a T. rex growth chart: [25] studies growth charts for several tyrannosaurid dinosaur species, including the tyrannosaurus rex (T. rex). There $n = 6$ T. rex observations with weights (kg) 29.9, 1761, 1807, 2984, 3230, 5040, and 5654 and corresponding ages

Listing 3.3
Gibbs sampling for the NFL concussions data.

```
1   # Load data
2
3   Y <- c(171, 152, 123, 199)
4   n <- 4
5   N <- 256
6
7   # Create an empty matrix for the MCMC samples
8
9   S                   <- 25000
10  samples             <- matrix(NA,S,5)
11  colnames(samples) <- c("lam1","lam2","lam2","lam4","gamma")
12
13  # Initial values
14
15  lambda <- log(Y/N)
16  gamma  <- 1/mean(lambda)
17
18  # priors: lambda[i]|gamma ~ Gamma(1,gamma), gamma ~ InvG(a,b)
19
20  a <- 0.1
21  b <- 0.1
22
23  # Gibbs sampling
24
25  for(s in 1:S){
26    for(i in 1:n){
27      lambda[i] <- rgamma(1,Y[i]+1,N+gamma)
28    }
29    gamma        <- rgamma(1,4+a,sum(lambda)+b)
30    samples[s,] <- c(lambda,gamma)
31  }
32
33  boxplot(samples[,1:4],outline=FALSE,
34          ylab=expression(lambda),names=2012:2015)
35  plot(samples[,5],type="l",xlab="Iteration",
36       ylab=expression(gamma))
37
38  # Posterior mean, median and credible interval
39  apply(samples,2,mean)
40  apply(samples,2,quantile,c(0.025,0.500,0.975))
```

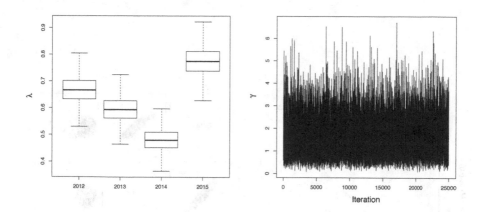

FIGURE 3.6
MCMC analysis of the NFL concussions data. The left panel plots the posterior distribution of the concussion rate λ_i for each year, and the right panel is a trace plot of the hyperparameter γ.

(years) 2, 15, 14, 16, 18, 22, and 28, plotted in Figure 3.7 (top left). Non-linear models are likely more appropriate for growth curve data (Section 6.3), but we fit a linear model for illustration. We assume that $Y_i = \beta_1 + x_i\beta_2 + \varepsilon_i$ where Y_i is the weight of dinosaur i, x_i is the age, β_1 and β_2 are the regression intercept and slope, respectively, and $\varepsilon_i \overset{iid}{\sim} \text{Normal}(0, \sigma^2)$. We select uninformative priors $\beta_j \overset{indep}{\sim} \text{Normal}(0, \tau)$ and $\sigma^2 \sim \text{InvGamma}(a, b)$ with $\tau = 10{,}000^2$ and $a = b = 0.1$. Figure 3.7 (top right) plots the posterior distribution of the three parameters $\boldsymbol{\theta} = (\beta_1, \beta_2, \sigma)$. The error standard deviation σ is not strongly dependent on the other parameters, but the regression coefficients $\boldsymbol{\beta} = (\beta_1, \beta_2)$ have correlation $\text{Cor}(\beta_1, \beta_2) = -0.91$.

Gibbs sampling has trouble when parameters have strong posterior dependence. To see this, consider the hypothetical example plotted in Figure 3.8. The two parameters $\boldsymbol{\beta} = (\beta_1, \beta_2)$ follow a bivariate normal posterior with means equal zero, variances equal 1, and correlation equal -0.98. The initial values are $\boldsymbol{\beta}^{(0)} = (0, -3)$. The first step is to update β_1 with $\beta_2 = -3$. This corresponds to a draw along the bottom row of Figure 3.8, and so the sample is $\beta_1 \approx 3$. The next update is from the conditional distribution of β_2 given $\beta_1 \approx 3$. Since the two variables are negatively correlated, with $\beta_1 \approx 3$ we must have $\beta_2 \approx -3$, and so β_2 is only slightly changed from its initial value. These small updates continue for the next four iterations and $\boldsymbol{\beta}$ stays in the lower-right quadrant. This toy example illustrates that with strong posterior dependence between parameters, the one-at-a-time Gibbs sampler will slowly traverse the parameter space, leading to poor convergence.

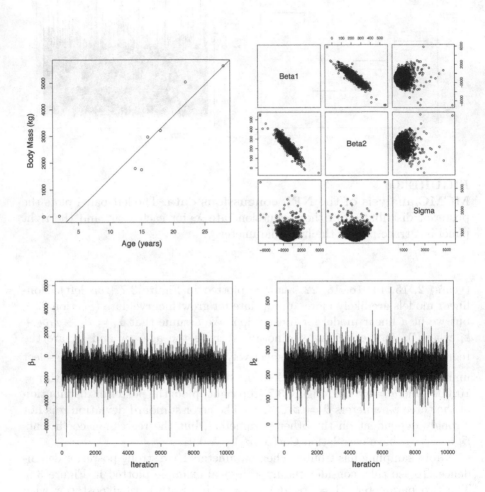

FIGURE 3.7
Analysis of the T. rex growth chart. The data are plotted in the top left panel and samples from the joint posterior of the three model parameters are plotted in the top right. The second row gives trace plots of β_1 and β_2.

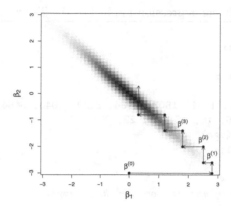

FIGURE 3.8
Gibbs sampling for correlated parameters. The background color represents the posterior PDF (black is high, white is low) of a hypothetical bivariate posterior for $\boldsymbol{\beta} = (\beta_1, \beta_2)$, and dots represent the starting value $\boldsymbol{\beta}^{(0)}$ and five hypothetical Gibbs sampling updates $\boldsymbol{\beta}^{(1)}, ..., \boldsymbol{\beta}^{(5)}$.

One way to improve convergence is to update dependent parameters in blocks. Returning to the T. rex example, since β_1 and β_2 are correlated with each other but not σ^2 (Figure 3.7), we might set blocks $\boldsymbol{\theta}_1 = \boldsymbol{\beta}$ and $\theta_2 = \sigma^2$ and apply Algorithm 1 by alternating between sampling from $\boldsymbol{\theta}_1|\theta_2, \mathbf{Y}$ and $\theta_2|\boldsymbol{\theta}_1, \mathbf{Y}$. Since $\boldsymbol{\beta}$ is drawn from its joint posterior, it will not get stuck in a corner as it might in one-at-a-time sampling, and thus convergence is improved.

To derive the full conditional distributions of the blocks and MCMC code we use matrix notation. Let the data vector be $\mathbf{Y} = (Y_1, ..., Y_n)^T$ and $\mathbf{X}$ be $n \times 2$ covariate matrix with the ith row equal to $(1, x_i)$. The linear regression model can be written $\mathbf{Y}|\boldsymbol{\beta}, \sigma^2 \sim \text{Normal}(\mathbf{X}\boldsymbol{\beta}, \sigma^2 \mathbf{I}_n)$. The full conditional distributions (derived in Appendix A.3 and discussed in Section 2.1.6) are

$$\boldsymbol{\beta}|\sigma^2, \mathbf{Y} \sim \text{Normal}\left(\mathbf{P}^{-1}\mathbf{W}, \mathbf{P}^{-1}\right)$$
$$\sigma^2|\boldsymbol{\beta}, \mathbf{Y} \sim \text{InvGamma}\left(n/2 + a, (\mathbf{Y} - \mathbf{X}\boldsymbol{\beta})^T(\mathbf{Y} - \mathbf{X}\boldsymbol{\beta})/2 + b\right)$$

where $\mathbf{P} = \mathbf{X}^T\mathbf{X}/\sigma^2 + \mathbf{I}/\tau$ and $\mathbf{W} = \mathbf{X}^T\mathbf{Y}/\sigma^2$. Code to implement this model is given in Listing 3.4. Figure 3.7 (bottom row) shows excellent convergence. If there were more than one covariate, then $\mathbf{X}$ would have additional columns and $\boldsymbol{\beta}$ would have additional elements, but the full conditionals and Gibbs sampling steps would be unchanged.

Listing 3.4
Gibbs sampling for linear regression applied to the T-rex data.

```
1
2   library(mvtnorm)
3
4   # Load T-Rex data
5
6    mass <- c(29.9, 1761, 1807, 2984, 3230, 5040, 5654)
7    age <- c(2, 15, 14, 16, 18, 22, 28)
8    n   <- length(age)
9    X   <- cbind(1,age)
10   Y   <- mass
11
12  # Create an empty matrix for the MCMC samples
13
14   S                  <- 10000
15   samples            <- matrix(NA,S,3)
16   colnames(samples) <- c("Beta1","Beta2","Sigma")
17
18  # Initial values
19
20   beta <- lm(mass~age)$coef
21   sig2 <- var(lm(mass~age)$residuals)
22
23  # priors: beta ~ N(0,tau I_2), sigma^2 ~ InvG(a,b)
24
25   tau    <- 10000^2
26   a      <- 0.1
27   b      <- 0.1
28
29  # Blocked Gibbs sampling
30
31   V      <- diag(2)/tau
32   tXX    <- t(X)%*%X
33   tXY    <- t(X)%*%Y
34
35   for(s in 1:S){
36     P    <- tXX/sig2 + V
37     W    <- tXY/sig2
38     beta <- rmvnorm(1,solve(P)%*%W,solve(P))
39     beta <- as.vector(beta)
40
41     A    <- n/2 + a
42     B    <- sum((Y-X%*%beta)^2)/2 + b
43     sig2 <- 1/rgamma(1,A,B)
44
45     samples[s,]<-c(beta,sqrt(sig2))
46   }
47
48   pairs(samples)
```

3.2.2 Metropolis–Hastings (MH) sampling

Each step in Gibbs sampling requires taking a sample from the full conditional distribution of one parameter (or block of parameters) conditional on all other parameters. In the examples in the previous section, all priors were conditionally conjugate, and so we were able to show that the full conditionals were members of familiar families of distributions and so sampling was straightforward. However, it is not always possible to specify a conditionally conjugate prior.

For example, returning the NFL concussions data (Figure 3.9), say we select the model

$$Y_i|\boldsymbol{\beta} \sim \text{Poisson}[N \exp(\beta_1 + \beta_2 i)] \tag{3.13}$$

for year $i = 1, ..., 4$ so that the log concussion rate changes linearly in time. The likelihood is then

$$f(\mathbf{Y}|\boldsymbol{\beta}) \propto \prod_{i=1}^{4} \exp[-N \exp(\beta_1 + \beta_2 i)] \exp[Y_i(\beta_1 + \beta_2 i)]. \tag{3.14}$$

The parameters appear in the exponential within another exponential, and there is no well-known family of distributions for $\boldsymbol{\beta}$ with this feature. Therefore, the posterior will not belong to a known distribution, and it is not clear how to directly sample from the posterior.

MH sampling [41] replaces draws from the exact full conditional distribution with a draw from a candidate distribution followed by an accept/reject step. That is, the Gibbs update of β_1 is a sample $\beta_1^{(s)}|\beta_2^{(s-1)}, \mathbf{Y}$, whereas an MH makes a candidate draw $\beta_1^* \sim q(\beta_1|\beta_1^{(s-1)})$ that is conditioned on the current value of β_1 (and potentially β_2 and/or $\mathbf{Y}$). Of course, the candidate cannot be blindly accepted because the candidate distribution may not be related to the posterior distribution. To correct for this, the candidate is accepted with probability $\min\{1, R\}$ where R is the ratio

$$R = \frac{p(\beta_1^*|\beta_2, \mathbf{Y})}{p(\beta_1^{(s-1)}|\beta_2, \mathbf{Y})} \frac{q(\beta_1^{(s-1)}|\beta_1^*)}{q(\beta_1^*|\beta_1^{(s-1)})}. \tag{3.15}$$

Equivalently, the accept/reject step generates $U \sim \text{Uniform}(0, 1)$ and accepts the candidate if $U < R$. Algorithm 2 formally describes the MH sampler, which can be justified following the steps outlined in Appendix A.3 for Gibbs sampling.

The acceptance ratio R depends on the ratio of the posteriors of the candidate and current values. For updating parameter θ_j, terms in the likelihood or prior that do not include θ_j cancel and can thus be ignored. Crucially, this includes the often intractable normalizing constant $m(\mathbf{Y})$. Other terms can cancel as well, for example, the entire likelihood for the example in Listing 3.3 did not depend on one parameter (γ) and would cancel in the posterior ratio for this parameter.

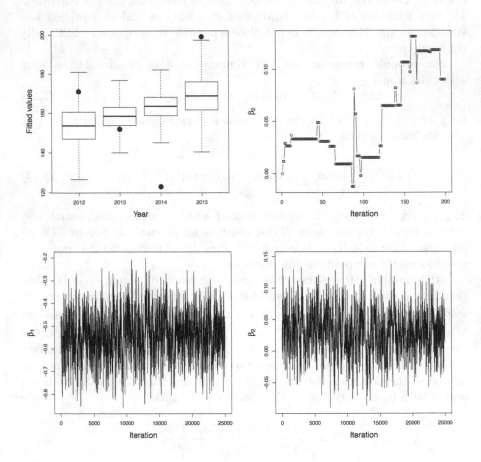

FIGURE 3.9
Poisson regression for the NFL concussions data. The first panel plots
the posterior distribution of the mean values $N\exp(\beta_1 + \beta_2 i)$ for each year
(boxplots) and the observed number of concussions, Y_i (points); the remaining
three panels are trace plots of the regression coefficients β_1 and β_2.

Algorithm 2 Metropolis–Hastings Sampling

1: Initialize $\boldsymbol{\theta}^{(0)} = (\theta_1^{(0)}, ..., \theta_p^{(0)})$
2: **for** $s = 1, ..., S$ **do**
3:      **for** $j = 1, ..., p$ **do**
4:          sample $\theta_j^* \sim q_j(\theta_j | \theta_j^{(s-1)})$
5:          set $\boldsymbol{\theta}^* = \left(\theta_1^{(s)}, ..., \theta_{j-1}^{(s)}, \theta_j^*, \theta_{j+1}^{(s-1)}, ..., \theta_p^{(s-1)} \right)$
6:          set $R = \dfrac{f(\mathbf{Y}|\boldsymbol{\theta}^*)\pi(\boldsymbol{\theta}^*)}{f\left(\mathbf{Y}|\boldsymbol{\theta}^{(s-1)}\right)\pi\left(\boldsymbol{\theta}^{(s-1)}\right)} \cdot \dfrac{q_j\left(\theta_j^{(s-1)}|\theta_j^*\right)}{q_j\left(\theta_j^*|\theta_j^{(s-1)}\right)}$
7:          sample $U \sim \text{Uniform}(0,1)$
8:          **if** $U < R$ **then**
9:             set $\theta_j^{(s)} = \theta_j^*$
10:          **else**
11:             set $\theta_j^{(s)} = \theta_j^{(s-1)}$
12:          **end if**
13:      **end for**
14: **end for**

The MH sampler can be applied more generally than Gibbs sampling, but at the cost of having to select and tune a candidate distribution for each parameter (or block of parameters). A common choice is a random-walk Gaussian candidate distribution

$$\theta_j^* | \theta_j^{(s-1)} \sim \text{Normal}(\theta_j^{(s-1)}, c_j^2) \tag{3.16}$$

where c_j is the candidate standard deviation. This is called a random-walk proposal distribution because it simply adds Gaussian jitter to the current state of the chain. A Gaussian candidate distribution can be used for any continuous parameter, even for those without a Gaussian prior. This is not always ideal. For example, a Gaussian candidate for a variance parameter may propose negative values which will automatically be discarded because the prior density and thus acceptance probability will be zero. Alternatively, a parameter with bounded support (e.g., $\sigma > 0$) can be transformed to have support over the whole real line ($\theta = \log(\sigma)$) and a Gaussian can be used in this transformed space. A Gaussian candidate is not appropriate for a discrete parameter, however, because Gaussian draw would almost surely not be in the support of the discrete parameter's prior.

The standard deviation of the candidate distribution, c_j, is a tuning parameter. A rule of thumb is to tune the algorithm so that it accepts 30-50% of the candidates. It is hard to know which value of c_j will lead to this acceptance rate before starting the sampling, and so it is common to adapt c_j based on batches of samples during the burn-in. For example, if less than 30% of the last 100 candidates were accepted then you might decrease c_j by 20%, and if more than 50% of the last 100 candidates were accepted then you might increase c_j

by 20%. However, once the burn-in has completed, you must fix the candidate distribution unless you consider more advanced methods (Appendix A.4).

Benefits of the random-walk candidate distribution are that it does not require knowledge about the form of the posterior, and that the acceptance ratio simplifies. If the candidate PDF q is Gaussian,

$$q(\theta_j^{(s-1)}|\theta_j^*) = q_j(\theta_j^*|\theta_j^{(s-1)}) = \frac{1}{\sqrt{2\pi}c_j} \exp\left(-\frac{(\theta_j^{(s-1)} - \theta_j^*)^2}{2c_j^2}\right), \qquad (3.17)$$

and so the ratio of candidate distributions in the MH acceptance probabilities cancels. This is called a symmetric candidate distribution, and the MH algorithm reduces to the Metropolis algorithm [55] in Algorithm 3 if the candidate distribution is symmetric.

Algorithm 3 Metropolis Sampling

1: Initialize $\boldsymbol{\theta}^{(0)} = (\theta_1^{(0)}, ..., \theta_p^{(0)})$
2: **for** $s = 1, ..., S$ **do**
3:     **for** $j = 1, ..., p$ **do**
4:         sample $\theta_j^* \sim q_j(\theta_j|\theta_j^{(s-1)})$ for symmetric q_j
5:         set $\boldsymbol{\theta}^* = \left(\theta_1^{(s)}, ..., \theta_{j-1}^{(s)}, \theta_j^*, \theta_{j+1}^{(s-1)}, ..., \theta_p^{(s-1)}\right)$
6:         set $R = \dfrac{f(\mathbf{Y}|\boldsymbol{\theta}^*)\pi(\boldsymbol{\theta}^*)}{f(\mathbf{Y}|\boldsymbol{\theta}^{(s-1)})\pi(\boldsymbol{\theta}^{(s-1)})}$
7:         sample $U \sim \text{Uniform}(0, 1)$
8:         **if** $U < R$ **then**
9:             set $\theta_j^{(s)} = \theta_j^*$
10:        **else**
11:           set $\theta_j^{(s)} = \theta_j^{(s-1)}$
12:        **end if**
13:     **end for**
14: **end for**

Listing 3.5 gives R code for the Poisson regression model of NFL concussions. The Gaussian candidate distributions were set to have standard deviation 0.1. Since the parameters appear in the exponential function in the mean count, adding normal jitter with standard deviation 0.1 is similar to a 10% change in the mean counts, which leads to acceptance probability of 0.42 for β_1 and 0.18 for β_2. Further tuning might reduce the candidate standard deviation for β_2 to increase the acceptance probability, but the trace plots in Figure 3.9 show good convergence. The top right panel of Figure 3.9 zooms in on the first few samples of the Metropolis algorithm and shows how β_2 stays constant for several iterations before jumping to a new value when a candidate is accepted.

For the accept/reject step in Lines 40–41 of Listing 3.5 it is important to perform computation on the log scale. For larger problems, the ratio of

Listing 3.5
Metropolis sampling for the NFL concussions data.

```
1   # Load data
2
3   Y <- c(171, 152, 123, 199)
4   t <- 1:4
5   n <- 4
6   N <- 256
7
8   # Create an empty matrix for the MCMC samples
9
10  S                  <- 25000
11  samples            <- matrix(NA,S,2)
12  colnames(samples)  <- c("beta1","beta2")
13  fitted             <- matrix(NA,S,4)
14
15  # Initial values
16
17  beta   <- c(log(mean(Y/N)),0)
18
19  # priors: beta[j] ~ N(0,tau^2)
20
21  tau    <- 10
22
23  # Prep for Metropolis sampling
24
25  log_post <- function(Y,N,t,beta,tau){
26    mn    <- N*exp(beta[1]+beta[2]*t)
27    like  <- sum(dpois(Y,mn,log=TRUE))
28    prior <- sum(dnorm(beta,0,tau,log=TRUE))
29    post  <- like + prior
30  return(post)}
31
32  can_sd <- rep(0.1,2)
33
34  # Metropolis sampling
35
36  for(s in 1:S){
37    for(j in 1:2){
38      can    <- beta
39      can[j] <- rnorm(1,beta[j],can_sd[j])
40      logR   <- log_post(Y,N,t,can,tau)-log_post(Y,N,t,beta,tau)
41      if(log(runif(1))<logR){
42        beta <- can
43      }
44    }
45    samples[s,] <- beta
46    fitted[s,] <- N*exp(beta[1]+beta[2]*t)
47  }
48
49  boxplot(fitted,outline=FALSE,ylim=range(Y),
50          xlab="Year",ylab="Fitted values",names=2012:2015)
51  points(Y,pch=19)
```

posteriors is numerically unstable because a small value in the denominator can cause the ratio to be numerically infinite. Therefore, the ratio is replaced by the difference of the log scale. The original rejection step is to reject if $U < f(\boldsymbol{\theta}^*|\mathbf{Y})/f(\boldsymbol{\theta}^{(s-1)}|\mathbf{Y})$. Taking the log of each side gives the equivalent inequality $\log(U) < \log[f(\boldsymbol{\theta}^*|\mathbf{Y})] - \log[f(\boldsymbol{\theta}^{(s-1)}|\mathbf{Y})]$.

Random-walk Gaussian distributions can also be used for block updates. As with Gibbs, updating dependent parameters simultaneously can improve convergence. Say the p parameters are partitioned into q blocks, $\boldsymbol{\theta} = (\boldsymbol{\theta}_1, ..., \boldsymbol{\theta}_q)$, and the candidate distribution for block j is $\boldsymbol{\theta}_j^* \sim$ Normal$(\boldsymbol{\theta}^{(s-1)}, c_j \mathbf{V}_j)$. As before, the scalar c_j must be tuned to give a reasonable acceptance probability. The matrix $\mathbf{V}_j$ must also be tuned, which is difficult because it involves the variance of each parameter and the correlation between each pair of parameters, and all of these elements contribute to the single acceptance probability. A reasonable choice is to set $\mathbf{V}_j$ to the posterior covariance of $\boldsymbol{\theta}_j$, but unfortunately this is unknown. A common remedy is to use a short burn-in period and set $\mathbf{V}_j$ to the sample covariance of the burn-in samples, and then tune the scalar c_j to give reasonable acceptance. The proposal distribution can also be adapted based on prior MCMC samples (Appendix A.4)

A drawback of the random walk candidate distribution is that it may be suboptimal if it does not closely approximate the posterior. Using additional information can improve convergence. For example, a Bayesian CLT (Section 3.1.3) approximation $\boldsymbol{\theta}|\mathbf{Y} \approx$ Normal$(\hat{\boldsymbol{\theta}}_{MAP}, \hat{\boldsymbol{\Sigma}}_{MAP})$ could be used to tune the sampler to match candidate to the posterior. Taking this to the extreme, if we can tune the candidate to be exactly the full conditional distribution, then the candidate distribution is proportional to the posterior, $q(\theta_j^*|\theta_j^{(s-1)}) \propto f(\mathbf{Y}|\boldsymbol{\theta}^*)\pi(\boldsymbol{\theta}^*)$, and the ratio in Line 6 of Algorithm 2 reduces to 1. This algorithm then cycles through the parameters, samples each from their full conditional distribution, and keeps all draws. This is the Gibbs sampling algorithm in Algorithm 1! This shows that Gibbs sampling is a special case of MH with careful selection of the candidate distributions. It also opens the door for mixing Gibbs and MH updates in the same algorithm because the algorithm can be framed as an MH sampler with some well-tuned candidates that lead to Gibbs steps.

Metropolis-within-Gibbs example: To illustrate this Metropolis-within-Gibbs sampling algorithm, we use logistic regression, which is an extension of linear regression for binary data. For binary data, $E(Y_i) = \text{Prob}(Y_i = 1)$, and thus we cannot directly model the mean as a linear combination of the covariates $\mathbf{X}_i = (1, X_{i2}..., X_{ip})$ because this linear combination might not be between zero and one. The linear regression model is thus modified to $Y_i|\boldsymbol{\beta} \sim$ Bernoulli(p_i), where $p_i = 1/[1 + \exp(-\eta_i)]$ for $\eta_i = \beta_1 + \sum_{j=2}^{p} \beta_j X_{ij}$, which ensures that the mean response is between zero and one. We select an uninformative prior for the intercept $\beta_1 \sim$ Normal$(0, 10^2)$, and for the slopes

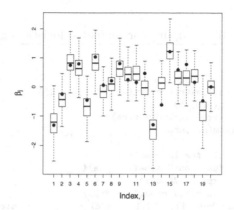

FIGURE 3.10
Logistic regression analysis of simulated data. The boxplots are the posterior distribution of the β_j and the points are the true values used to simulate the data.

with $j = 2, ..., p$ the prior is

$$\beta_j | \sigma^2 \sim \text{Normal}(0, \sigma^2) \text{ and } \sigma^2 \sim \text{InvGamma}(a, b). \tag{3.18}$$

The regression coefficients β_j appear in the likelihood inside a non-linear function, and so their normal prior is not conditionally conjugate. However, the prior variance σ^2 has conditionally conjugate inverse gamma prior and thus inverse gamma full conditional distribution. Therefore a Metropolis-within-Gibbs algorithm cycles through the β_j and updates them using Metropolis steps with Gaussian candidate distribution, and applies a Gibbs step to sample σ^2 from its inverse gamma full conditional distribution.

This model has $p + 1$ parameters and the algorithm has different types of updates and p tuning parameters. Due to its complexity, it should be interrogated before being used for a real data analysis. A common method for validating code is a simulation study. In a simulation study, we fix the values of the parameters and generate data using these values. The synthetic data are then analyzed and the posterior distributions are compared with the true values. Unlike a real data analysis, since the true values are known we can verify that the algorithm is able to recover the true values.

Listing 3.6 provides code to simulate the data, carry out the Metropolis-within-Gibbs algorithm, and summarize the results. The results are plotted in Figure 3.10. The true values of β_j used to generate the data are included by the posterior distribution for all but one or two coefficients, as expected. Therefore, it seems the algorithm is working well.

Listing 3.6
Metropolis-within-Gibbs sampling for simulated logistic regression data.

```
1   # Simulate data
2   n            <- 100
3   p            <- 20
4   X            <- cbind(1,matrix(rnorm(n*(p-1)),n,p-1))
5   beta_true <- rnorm(p,0,.5)
6   prob         <- 1/(1+exp(-X%*%beta_true))
7   Y            <- rbinom(n,1,prob)
8
9   # Function to compute the log posterior
10  log_post_beta <- function(Y,X,beta,sigma){
11      prob <- 1/(1+exp(-X%*%beta))
12      like <- sum(dbinom(Y,1,prob,log=TRUE))
13      prior <- dnorm(beta[1],0,10,log=TRUE) +         # Intercept
14               sum(dnorm(beta[-1],0,sigma,log=TRUE)) # Slopes
15  return(like+prior)}
16
17  # Create empty matrix for the MCMC samples
18  S                    <- 10000
19  samples              <- matrix(NA,S,p+1)
20
21  # Initial values and priors
22  beta  <- rep(0,p)
23  sigma <- 1
24  a     <- 0.1
25  b     <- 0.1
26  can_sd <- 0.1
27
28  for(s in 1:S){
29
30      # Metropolis for beta
31      for(j in 1:p){
32          can   <- beta
33          can[j] <- rnorm(1,beta[j],can_sd)
34          logR  <- log_post_beta(Y,X,can,sigma)-
35                   log_post_beta(Y,X,beta,sigma)
36          if(log(runif(1))<logR){
37              beta <- can
38          }
39      }
40
41      # Gibbs for sigma
42      sigma <- 1/sqrt(rgamma(1,(p-1)/2+a,sum(beta[-1]^2)/2+b))
43
44      samples[s,] <- c(beta,sigma)
45
46  }
47
48  boxplot(samples[,1:p],outline=FALSE,
49          xlab="Index, j",ylab=expression(beta[j]))
50  points(beta_true,pch=19,cex=1.25)
```

3.3   MCMC software options in R

Writing MCMC code step-by-step is great practice and the only way to truly understand the algorithms, and tailoring the code to a specific model can improve speed and stability for complex models. However, coding basic MCMC samplers becomes repetitive. For Gibbs sampling, you must derive and encode the full conditional distribution of each parameter, but there are only a few dozen known conjugacy pairs (e.g., Appendix A.2), and so writing a Gibbs sampler is simply a matter of looking up the correct full conditional distribution and entering this into R. If the parameter's full conditional distribution does not appear in the table of conjugacy pairs, you might use a Metropolis update with Gaussian random walk candidate distribution. Each Metropolis update consists of a Gaussian proposal followed by computing the acceptance ratio, which requires only tuning the candidate standard deviation and inserting the specific likelihood and prior chosen for the model into the acceptance ratio. But the tuning step is not model specific, and most models are built using only a few dozen distributions for the likelihood and prior (Appendix A.1), and so writing MH code becomes simply a matter of coding the correct distributions.

Fortunately, this process has been automated! There are several general-purpose MCMC packages that can be called from R, including Just Another Gibbs Sampler (JAGS), OpenBUGS ([79]), STAN ([15]), and NIMBLE ([22]), as well as libraries in python including pymc ([64]) and the SAS procedure PROC MCMC ([17]). These packages are not specific to a single model, rather they take a script specifying the likelihood and prior as input and use this information to construct an MCMC sampler for the specified model. Appendix A.5 provides code and a comparison of JAGS, OpenBUGS, STAN and NIMBLE on example datasets. The format of these packages are similar and so with a solid understanding of MCMC sampling to interpret the results, switching from one to another is fairly straightforward.

For the remainder of the book we will use JAGS to carry out MCMC sampling. Compared to other packages, JAGS is relatively easy to code and sufficiently fast for the size and complexity of models considered here. JAGS is a very general package that can be used to fit all the models discussed in this book and many more. Of course, for a specific application there may be more focused and thus more efficient code. For example, the BLR package in R surely performs better than JAGS for Bayesian linear regression. However, to avoid having to learn dozens of packages for different models we simply use JAGS throughout.

JAGS must first be downloaded from http://mcmc-jags.sourceforge.net/, and installed on your computer. The R package rjags must be downloaded and installed to communicate with JAGS from R. An MCMC analysis using JAGS has five steps:

(1) Specify the model as character string in R

(2) Upload the data and compile the model using the function `jags.model`

(3) Perform burn-in sampling using the function `update`

(4) Generate posterior samples using the function `coda.samples`

(5) Summarize the results using the functions `summarize` and `plot`.

Listing 3.7 contains the R code for simple linear regression applied to the T. rex growth chart data, and the output of the `plot` command in the last line is given in Figure 3.11. A few notes on this code:

- All of this code is run in R. You must install JAGS on your computer, but you never have to open JAGS because the `rjags` library passes data and results back and forth between R and JAGS.

- The model specification format resembles R code, but not all R commands are applicable in JAGS and some commands with the same syntax have different properties. For example, `dnorm` in JAGS specifies the normal distribution via its mean and precision (inverse variance) unlike the R command by the same name that uses mean and standard deviation. For a list of JAGS commands and their meaning see the user manual in [66].

- The symbol "~" indicates that a variable follows the distribution given to the right of the symbol, and deterministic operations are denoted by the left arrow "<-".

- In the model definition, the prior is placed on the precision `tau`, but each sample is converted to the standard deviation `sigma <- 1/sqrt(tau)` and samples of `sigma` are returned. That is, this line tells JAGS to compute $\sigma^{(s)} = 1/\sqrt{\tau^{(s)}}$ at each iteration and return the samples $\sigma^{(s)}$.

- The object `model` contains the data, the code to update each parameter, and the current state of each parameter in each chain.

- The initial values function can specify either random samples for initial values as is done here, or give specific values (e.g., `beta1 = 0`). If the `inits` argument in `jags.model` is omitted JAGS will set initial values automatically.

- The `update` function modifies the state of each parameter in each chain, but does not store intermediate samples.

- The `coda.samples` function does retain all MCMC samples, but only for those parameters listed in `variable.names`. The function outputs the samples into the object `samples` which are in the format used by the `coda` package that can be used to study convergence.

Listing 3.7
JAGS code for linear regression applied to the T-rex data.

```
1   library(rjags)
2   # Load T-Rex data
3   mass <- c(29.9, 1761, 1807, 2984, 3230, 5040, 5654)
4   age  <- c(2, 15, 14, 16, 18, 22, 28)
5   n    <- length(age)
6   data <- list(mass=mass,age=age,n=n)
7
8   # (1) Define the model as a string
9   model_string <- textConnection("model{
10    # Likelihood (dnorm uses a precision, not variance)
11    for(i in 1:n){
12      mass[i] ~ dnorm(beta1 + beta2*age[i],tau)
13    }
14    # Priors
15    tau   ~ dgamma(0.1, 0.1)
16    sigma <- 1/sqrt(tau)
17    beta1 ~ dnorm(0, 0.001)
18    beta2 ~ dnorm(0, 0.001)
19  }")
20
21  # (2) Load the data, specify initial values and compile the MCMC code
22  inits <- list(beta1=rnorm(1),beta2=rnorm(1),tau=rgamma(1,1))
23  model <- jags.model(model_string,data = data, inits=inits, n.chains=2)
24
25  # (3) Burn-in for 10000 samples
26  update(model, 10000, progress.bar="none")
27
28  # (4) Generate 20000 post-burn-in samples
29  params <- c("beta1","beta2","sigma")
30  samples <- coda.samples(model,
31             variable.names=params,
32             n.iter=20000, progress.bar="none")
33
34  # (5) Summarize the output
35  summary(samples)
36  1. Empirical mean and standard deviation for each variable,
37     plus standard error of the mean:
38
39          Mean      SD Naive SE Time-series SE
40  beta1    2.512  31.61  0.1580         0.1580
41  beta2   52.763  39.21  0.1961         0.3727
42  sigma 2792.738 1177.88 5.8894         9.7678
43
44  2. Quantiles for each variable:
45
46          2.5%     25%      50%     75%    97.5%
47  beta1 -59.53  -18.87    2.601   23.57    64.61
48  beta2 -21.36   25.71   51.531   78.34   134.17
49  sigma 1083.16 1997.85 2601.864 3361.14 5622.69
50
51  plot(samples)
```

- JAGS has built-in `plot` and `summary` functions that are used in Listing 3.7, but you can make your own plots by extracting the $S \times p$ matrix of samples for chain c, `samples[[c]]`. All the samples can be put in one $2S \times p$ matrix

```
samps_matrix <- rbind(samples[[1]],samples[[2]])
```

and, for example, the posterior quantiles could be computed as

```
apply(samps_matrix,2,quantile,
      c(0.025,0.250,0.500,0.750,0.975))
```

- The `summary` function combines the samples across the chains and gives the sample mean, standard deviation, and quantiles of the posterior samples.

- The posterior could be summarized using the posterior mean and 95% intervals of the marginal posterior for each parameter given by the `summarize` function (i.e., the "`Mean`" and "`2.5%`" and "`97.5%`" entries).

Listing 3.8 provides a second example of fitting the Poisson-gamma model to the NFL concussions data. The steps and code are very similar to the previous example, except that instead of using the built-in summary and plot function, Listing 3.8 extracts the samples from the `rjags` object `samples` and combines them into the $2S \times 4$ matrix of samples `samps`. This allows for more flexibility in the posterior summaries used to illustrate the results. For example, Listing 3.8 computes the 90% posterior intervals.

In the remainder of the book we will use JAGS for all MCMC coding. We will often show the model statement and summaries of the output, but we will not report the steps of loading the data, generating the samples, etc., because these blocks of code are virtually identical for all models.

3.4   Diagnosing and improving convergence

3.4.1   Selecting initial values

Theoretically, Gibbs and Metropolis–Hastings sampling should converge for any initial values, but in practice the choice of starting values is important. There are two schools of thought on initialization: select initial values close to the posterior mode and run one long chain or run several parallel chains with intentionally diffuse starting values to verify that the algorithm has converged.

Selecting good starting values and running one long chain has the advantage of shortening or eliminating the burn-in period. A common method to select initial values is the MLE or MAP estimates computed using numerical optimization, which is often easier to compute than MCMC sampling. A disadvantage of this approach is that it is hard to rule out the possibility that

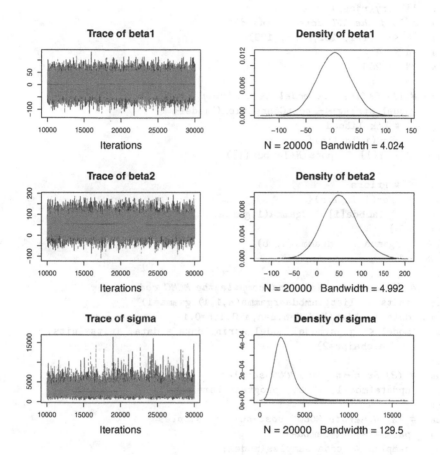

FIGURE 3.11
JAGS output for the T. rex analysis. This is the output of the `plot` function in JAGS for the linear regression of the T. rex growth chart data. The two lines with different shades of gray in the plots in the first columns are the samples from the two chains. The density plots on the right summarize the marginal distribution of each parameter combining samples across chains.

Listing 3.8
JAGS code for the NFL concussions data.

```
1   library(rjags)
2   # Load the NFL concussions data
3   Y <- c(171, 152, 123, 199)
4   n <- 4
5   N <- 256
6
7   # (1) Define the model as a string
8   model_string <- textConnection("model{
9     # Likelihood
10    for(i in 1:n){
11      Y[i] ~ dpois(N*lambda[i])
12    }
13    # Priors
14    for(i in 1:n){
15      lambda[i] ~ dgamma(1,gamma)
16    }
17    gamma   ~ dgamma(a, b)
18  }")
19
20  # (2) Load the data and compile the MCMC code
21  inits <- list(lambda=rgamma(n,1,1),gamma=1)
22  data <- list(Y=Y,N=N,n=n,a=0.1,b=0.1)
23  model <- jags.model(model_string,data = data, inits=inits,
        n.chains=2)
24
25  # (3) Burn-in for 10000 samples
26  update(model, 10000, progress.bar="none")
27
28  # (4) Generate 20000 post-burn-in samples
29  params <- c("lambda")
30  samples <- coda.samples(model,
31            variable.names=params,
32            n.iter=20000, progress.bar="none")
33
34  # (5) Compute 90% credible intervals
35  samps <- rbind(samples[[1]],samples[[2]]) #2S x 4 matrix of
        samples
36  apply(samps,2,quantile,c(0.05,0.95))
37
38         lambda[1] lambda[2] lambda[3] lambda[4]
39  2.5% 0.5722272 0.5035036 0.4005522 0.6717104
40  97.5% 0.7704071 0.6925751 0.5685348 0.8878783
```

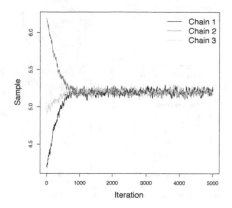

FIGURE 3.12
Convergence of three parallel chains. The three trace plots represent the samples for one parameter from three MCMC chains run in parallel. The chains converge around iteration 1,000.

the chain is stuck in a local mode and is completely missing the bulk of the posterior distribution.

On the other hand, starting multiple (typically 2–5) chains with diffuse starting values requires more burn-in samples, but if all chains give the same results then this provides evidence that the algorithm has properly converged. Figure 3.12 shows idealized convergence of three chains around iteration 500. MCMC is not easily parallelizable because of its sequential nature, but running several independent chains is one way to exploit parallel computing to improve Bayesian computation.

Both single and parallel chains have their relative merits, but given the progress in parallel computing environments it is preferable to run at least two chains. Care should be taken up front to ensure that the starting values of each chain are sufficiently spread across the posterior distribution so that convergence of the two chains necessarily represents strong evidence of convergence.

3.4.2 Convergence diagnostics

Verifying that the MCMC chains have converged and run long enough to sufficiently explore the posterior is often done informally by visual inspection of trace plots as in Figure 3.12. However, many formal diagnostics have been proposed to diagnose non-convergence. In this chapter, we focus on a few key diagnostics that are produced by JAGS via the coda package in R.

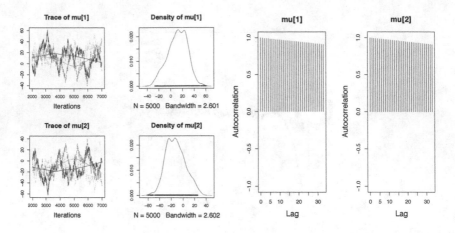

FIGURE 3.13
Convergence diagnostics for a toy example with poor convergence.
The left panel gives the trace plot for each parameter and each chain (distinguished by grayscale) and the right panel gives the autocorrelation functions for the first chain.

Throughout this section we use two toy examples for illustration:

$$\text{Poor convergence model} \quad Y|\boldsymbol{\mu} \sim \text{Poisson}[\exp(\mu_1 + \mu_2)]$$
$$\text{Good convergence model} \quad Y_1|\boldsymbol{\mu} \sim \text{Poisson}[\exp(\mu_1)]$$
$$Y_2|\boldsymbol{\mu} \sim \text{Poisson}[\exp(\mu_2)].$$

In both models the priors are $\mu_j \overset{iid}{\sim} \text{Normal}(0, 1000)$. In the first model the two parameters are unidentified and this leads to poor convergence, as shown in Listing 3.9 and Figure 3.13. In the second model the two parameters are related to separate observations and so both parameters are identified leading to good convergence as shown in Listing 3.10 and Figure 3.14. The first model is ridiculous and would never be fit to data, but provides a simple illustration of poor convergence. Of course, not all convergence problems are related to unidentified parameters, but this is a common source of problems.

The first diagnostic is to plot the trace plots and verify that the chains have all reached the same distribution and have mixed properly. Figure 3.14 (top left) provides a good example where the chains for both parameters look like a bar code, while Figure 3.13 (top left) is a problematic case where the chains are mixing slowly and likely need many more iterations to provide a good approximation to the posterior. In these small examples it is possible to plot all chains for all parameters, but for more complex models it may be necessary to inspect the chains for only a representative subset of the parameters.

Autocorrelation of the chains provides a numerical measure of how fast

Listing 3.9
Toy example with poor convergence.

```
1  # Define the model as a string
2  > model_string <- textConnection("model{
3  >    Y    ~ dpois(exp(mu[1]+mu[2]))
4  >    mu[1] ~ dnorm(0,0.001)
5  >    mu[2] ~ dnorm(0,0.001)
6  > }")
7
8  # Generate MCMC samples
9  > inits <- list(mu=rnorm(2,0,5))
10 > data <- list(Y=1)
11 > model <- jags.model(model_string,data = data,
12 >                          inits=inits, n.chains=3)
13
14 > update(model, 1000, progress.bar="none")
15 > samples <- coda.samples(model,
16 >             variable.names=c("mu"),
17 >             n.iter=5000, progress.bar="none")
18
19 ># Apply convergence diagnostics
20
21 > # Plots
22 > plot(samples)
23 > autocorr.plot(samples)
24
25 > # Statistics
26 > autocorr(samples[[1]],lag=1)
27 , , mu[1]
28          mu[1]      mu[2]
29 Lag 1 0.9948544 -0.9926385
30 , , mu[2]
31          mu[1]      mu[2]
32 Lag 1 -0.9960286 0.9947489
33
34 > effectiveSize(samples)
35    mu[1]    mu[2]
36 22.90147 22.71505
37
38 > gelman.diag(samples)
39      Point est. Upper C.I.
40 mu[1]      1.62       2.88
41 mu[2]      1.62       2.88
42
43 Multivariate psrf
44 1.48
45
46 > geweke.diag(samples[[1]])
47   mu[1]   mu[2]
48 -0.6555  0.6424
```

the chains are mixing. Ideally, the iterations would be independent of each other, but the Markovian nature of the sampler induces dependence between successive iterations. The lag-l autocorrelation of the chain for parameter θ_j is defined as

$$\rho_j(l) = \text{Cor}(\theta_j^{(s)}, \theta_j^{(s-l)}), \tag{3.19}$$

and the function $\rho_j(l)$ is called the autocorrelation function. The right panels of Figures 3.13 and 3.14 plot the sample autocorrelation functions for the two examples. Ideally we find that $\rho_j(l) \approx 0$ for all $l > 0$, but some correlation is expected. For example, the lag-1 autocorrelation for μ_1 in Figure 3.14 is around 0.4, but the chain converges nicely.

Rather than reporting the entire autocorrelation function for each parameter, the lag-1 autocorrelation is typically sufficient. Another common one-number summary of the entire function is the effective sample size (ESS). Recall that the sample mean of the MCMC samples is only an estimate of the true posterior mean, and we can quantify the uncertainty in this estimate with the standard error of the sample mean. If the samples are independent, then the standard error of the sample mean is $sd_j/\sqrt{S}$ where sd_j is the sample standard deviation of the draws for θ_j and S is the number of samples. This is labelled "naive SE" in the JAGS summary output as in Listing 3.7. However, this underestimates the uncertainty in the posterior-mean estimate if the samples have autocorrelation. It can be shown that the actual standard error accounting for autocorrelation is $sd_j/\sqrt{ESS_j}$ where

$$ESS_j = \frac{S}{1 + 2\sum_{l=1}^{\infty}\rho_j(l)} \leq S. \tag{3.20}$$

The standard error $sd_j/\sqrt{ESS_j}$ is labeled "Time-series SE" in the JAGS summary output. One way to determine if the number of samples is sufficient is to increase S until this standard error is acceptably low for all parameters. Another method is to increase S until the effective sample size is acceptably high for all parameters, say $ESS_j > 1000$.

Geweke's diagnostic ([36]) is used to detect non-convergence. It tests for convergence using a two-sample t-test to compare the mean of the chain between batches at the beginning versus the end of the sampler. The default in `coda` is to test that the mean is the same for the first 10% of the samples and the last 50% of the samples. Let $\bar{\theta}_{jb}$ and se_{jb} be the sample mean and standard error, respectively, for θ_j in batch $b = 1, 2$. The standard errors are computed accounting for autocorrelation (as in (3.20)) in the chains. The Geweke statistic is then

$$Z = \frac{\bar{\theta}_{j1} - \bar{\theta}_{j2}}{\sqrt{se_{j1}^2 + se_{j2}^2}}. \tag{3.21}$$

Under the null hypothesis that the means are the same for batches (and assuming each batch has large ESS), Z follows a standard normal distribution and so $|Z| > 2$ is cause for concern. In the two examples in Listings 3.9 and 3.10, $|Z|$ is less than one and so this statistic does not detect non-convergence.

Listing 3.10
Toy example with good convergence.

```
1   # Define the model as a string
2   > model_string <- textConnection("model{
3   >   Y1    ~ dpois(exp(mu[1]))
4   >   Y2    ~ dpois(exp(mu[2]))
5   >   mu[1] ~ dnorm(0,0.001)
6   >   mu[2] ~ dnorm(0,0.001)
7   > }")
8
9   # Generate MCMC samples
10  > inits <- list(mu=rnorm(2,0,5))
11  > data <- list(Y1=1,Y2=10)
12  > model <- jags.model(model_string,data = data,
13  >                         inits=inits, n.chains=3)
14
15  > update(model, 1000, progress.bar="none")
16  > samples <- coda.samples(model,
17  >             variable.names=c("mu"),
18  >             n.iter=5000, progress.bar="none")
19
20  ># Apply convergence diagnostics
21
22  > # Plots
23  > plot(samples)
24  > autocorr.plot(samples)
25
26  > # Statistics
27  > autocorr(samples[[1]],lag=1)
28  , , mu[1]
29         mu[1]      mu[2]
30  Lag 1 0.359733 0.02112005
31  , , mu[2]
32           mu[1]     mu[2]
33  Lag 1 0.002213494 0.2776712
34
35  >  effectiveSize(samples)
36    mu[1]    mu[2]
37  6494.326 8227.748
38  >  gelman.diag(samples)
39
40       Point est. Upper C.I.
41  mu[1]         1          1
42  mu[2]         1          1
43  Multivariate psrf
44  1
45
46  > geweke.diag(samples[[1]])
47    mu[1]    mu[2]
48  -0.5217 -0.2353
```

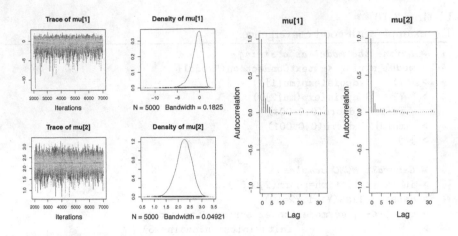

FIGURE 3.14
Convergence diagnostics for a toy example with good convergence.
The left panel gives the trace plot for each parameter and each chain (distinguished by grayscale) and the right panel gives the autocorrelation functions for the first chain.

Geweke's statistic uses only one chain. A multi-chain extension is the Gelman–Rubin statistic ([32]). The Gelman–Rubin statistic measures agreement in the means between C chains. It is essentially an ANOVA test of whether the chains have the same mean, but the statistic is scaled so that 1.0 indicates perfect convergence and values above 1.1 are questionable. The statistic is

$$R_j = \frac{\frac{S-1}{S}W + \left(\frac{1}{S} + \frac{1}{SC}\right)B}{W} \tag{3.22}$$

where B is S times the variance of the C MCMC sample means for θ_j and W is the average of the C MCMC sample variances for θ_j. The `coda` package plots the Gelman–Rubin statistic as a function of the iteration, and so when the statistic reaches one this indicates good mixing among the parallel chains and that the chains have possibly reached their stationary distribution. The Gelman–Rubin statistics clearly distinguish between the examples in Listing 3.9 and 3.10; $R_j = 1.62$ in the poor-convergence case and $R_j = 1.00$ for the good-convergence case.

3.4.3   Improving convergence

Armed with these diagnostic tools, if the user honestly inspects the sample output, poor convergence can usually be detected. This leaves the more challenging problem of improving convergence! There is no single step that always resolves convergence problems, but the list below offers some suggestions:

1. **Increase the number of iterations**: Theoretically MCMC algorithms should converge and cover the posterior distribution as S goes to infinity. Of course, time limits the number of samples that can be generated, but in some cases increasing S can be faster than searching for other improvements. Nonetheless, this is not the most satisfying solution to poor convergence, specifically when the chains move slowly due to high autocorrelation.

2. **Tune the MH candidate distribution**: The acceptance probability can be tuned during burn-in to be around 30-50%, but as the chain evolves the acceptance probability can change. If groups of parameters have strong cross-correlation then updating them in blocks can dramatically improve convergence. Block steps must be carefully tuned however so the candidate correlation matrix approximates the posterior correlation matrix.

3. **Improve initial values**: Often a sampler works well once it finds the posterior distribution, but the posterior PDF may be too flat away from the mode for the sampler to find its center. One way to improve initial values is to use the maximum likelihood estimates, $\boldsymbol{\theta}^{(0)} = \hat{\boldsymbol{\theta}}_{MLE}$. Another option is to use simulated annealing during burn-in. In simulated annealing, the MH acceptance rate is raised to the power $T_s \in (0,1]$, i.e., R^{T_s}, where the temperature T_s increases to one throughout the burn-in, e.g., $T_s = s/B$ for $s < B$ and $T_s = 1$ for $s > B$, where B is the number of burn-in samples. The intuition behind this modification is that by raising the acceptance ratio to a power the algorithm is more likely to make big jumps during the burn-in and will slowly settle down in the heart of the posterior as the burn-in period ends.

4. **Use a more advanced algorithm**: Appendix A.4 provides several advanced algorithms that can be used for a particularly vexing problem. For example, Hamiltonian Monte Carlo (HMC) is a Metropolis sampler that uses the gradient of the posterior to intelligently propose candidates and adaptive Metropolis estimates allows the proposal distribution to evolve across iterations.

5. **Simplify the model**: Poor convergence often occurs when models are too complex to be fit with the data at hand. Overly complex models often have unidentified parameters that cannot be estimated. For example, in a model with $Y_i \stackrel{iid}{\sim} \text{Normal}(\theta_1 + \theta_2, \sigma^2)$, the two mean parameters are not identified, meaning that there are infinitely many combinations of the parameters that give the same likelihood (e.g., $\boldsymbol{\theta} = (-10, 10)$ or $\boldsymbol{\theta} = (10, -10)$ give the same mean). Of course, no one would fit such a blatantly unidentified model, but in complex models with dozens of parameters, unidentifiability can be hard to detect. Convergence can be improved by sim-

plifying the model by eliminating covariates, removing non-linear terms, reducing the covariance structure to independence, etc.

6. **Pick more informative priors**: Selecting more informative priors has a similar effect to simplifying the model. For example, even in the silly model $Y_i \overset{iid}{\sim} \text{Normal}(\theta_1 + \theta_2, \sigma^2)$, the MCMC algorithm would likely converge quickly with priors $\theta_1 \sim \text{Normal}(-3, 1)$ and $\theta_2 \sim \text{Normal}(3, 1)$. Even a weakly informative prior can stabilize the posterior and improve convergence. As an extreme example, an empirical Bayesian prior (Chapter 2) fixes nuisance parameters at their MAP estimates, which can dramatically improve convergence at the expense of suppressing some uncertainty.

7. **Run a simulation study**: Determining whether a chain has converged can be frustrating for a real data analysis where the true values of the parameters are unknown. Simulating data from the model and then fitting the MCMC algorithm (as in Listing 3.6) for different sample sizes and parameter values can shed light on the properties of the algorithm and build trust that the sampler is producing reliable output.

3.4.4   Dealing with large datasets

Big data poses computational challenges to all statistical methods, but its effects are acutely felt by MCMC because MCMC requires thousands of passes through the data. Fortunately, recent years have seen a surge in Bayesian computing methods to handle massive datasets. Likely the next few years will see further developments in this area, but below we outline some of the most useful current approaches.

1. **MAP estimation**: MCMC is the gold standard for Bayesian computing because it returns estimates of the entire joint distribution of all of the parameters. However, if the scope of the analysis is limited to prediction, then it is much faster to simply compute the MAP estimate and ignore uncertainty in parameter estimation.

2. **Gaussian approximation**: The Bayesian CLT states that the posterior is approximately normal if the sample size is large. Therefore, running a long MCMC chain for, say, a logistic regression with $n = 1,000,000$ observations, $p = 50$ covariates and flat priors is unjustifiable; it is much faster and equally accurate to approximate the posterior as Gaussian centered on the MAP estimate with covariance determined by the information matrix. This computation can often be carried out using MLE software, and thus the posterior will be very similar to the approximate sampling distribution of the MLE, but the interpretation of uncertainty remains Bayesian.

3. **Variational Bayesian computing**: Variational Bayesian approximations can be even faster to compute than evoking the Bayesian CLT if you target only the marginal distribution of each parameter. That is, if you assume the posterior (not prior) is independent across parameters and use the best approximation to the true posterior that has this form, the posterior can be approximated without computing the joint distribution. This of course raise questions about properly accounting for uncertainty, but does lead to impressive computational savings.

4. **Parallel computing**: MCMC is inherently sequential, i.e., you generally cannot update the second parameter until after updating the first. In special cases, steps of the MCMC routine can be performed simultaneously in parallel on different cores of a CPU or GPU cluster. For example, if parameters are (conditionally) independent then they can be updated in parallel. Alternatively, if the data are independent and the likelihood factors as the product of n terms, then likelihood calculations within each MCMC step can be done in parallel, e.g., n_1 terms computed on core 1, n_2 terms computed on core 2, etc. While parallelization can theoretically have huge benefits, the overhead associated with passing data across cores can dampen the benefits of using multiple cores unless parallelization is done carefully.

5. **Divide and conquer**: Divide and conquer methods provide an approximation to the posterior that is embarrassingly parallelizable. Say we partition the data into T batches $\mathbf{Y}_1, ..., \mathbf{Y}_T$ that are independent given the model parameters. Then the posterior can be written

$$p(\boldsymbol{\theta}|\mathbf{Y}) \propto f(\mathbf{Y}|\boldsymbol{\theta})\pi(\boldsymbol{\theta}) = \prod_{t=1}^{T}\left[f(\mathbf{Y}_t|\boldsymbol{\theta})\pi(\boldsymbol{\theta})^{1/T}\right]. \tag{3.23}$$

This decomposition suggests fitting T separate Bayesian analyses (in parallel) with analysis t using data $\mathbf{Y}_t$ and prior $\pi(\boldsymbol{\theta})^{1/T}$ and then combining the results. Each of the T analyses uses the original prior $\pi(\boldsymbol{\theta})$ raised to the $1/T$ power to spread the prior information across batches, e.g., if $\theta \sim \text{Normal}(0, \tau^2)$ then the prior to the $1/T$ power is $\text{Normal}(0, T\tau^2)$. Scott et al. [75] discuss several ways to combine the T posterior distributions. The simplest method is to assume the posteriors are approximately Gaussian and thus analysis t returns $\boldsymbol{\theta}|\mathbf{Y}_t \approx \text{Normal}(\mathbf{M}_t, \mathbf{V}_t)$. In this case, the posteriors combine as

$$p(\boldsymbol{\theta}|\mathbf{Y}) \approx \text{Normal}(\mathbf{V}\sum_{t=1}^{T}\mathbf{V}_t^{-1}\mathbf{M}_t, \mathbf{V}) \tag{3.24}$$

where $\mathbf{V}^{-1} = \sum_{t=1}^{T}\mathbf{V}_t^{-1}$.

3.5 Exercises

1. Give an advantage and a disadvantage of the following methods:

    (a) Maximum a posteriori estimation
    (b) Numerical integration
    (c) Bayesian central limit theorem
    (d) Gibbs sampling
    (e) Metropolis–Hastings sampling

2. Assume that $Y_i|\mu \overset{indep}{\sim} \text{Normal}(\mu, \sigma_i^2)$ for $i \in \{1, ..., n\}$, with σ_i known and improper prior distribution $\pi(\mu) = 1$ for all μ.

    (a) Give a formula for the MAP estimator for μ.
    (b) We observe $n = 3$, $Y_1 = 12$, $Y_2 = 10$, $Y_3 = 22$, $\sigma_1 = \sigma_2 = 3$ and $\sigma_3 = 10$, compute the MAP estimate of μ.
    (c) Use numerical integration to compute the posterior mean of μ.
    (d) Plot the posterior distribution of μ and indicate the MAP and the posterior mean estimates on the plot.

3. Assume $Y_i|\lambda \overset{indep}{\sim} \text{Poisson}(N_i\lambda)$ for $i = 1, ..., n$.

    (a) Identify a conjugate prior for λ and derive the posterior that follows from this prior.
    (b) Using the prior $\lambda \sim \text{Uniform}(0, 20)$, derive the MAP estimator of λ.
    (c) Using the prior $\lambda \sim \text{Uniform}(0, 20)$, plot the posterior on a grid of λ assuming $n = 2$, $N_1 = 50$, $N_2 = 100$, $Y_1 = 12$, and $Y_2 = 25$ and show that the MAP estimate is indeed the maximizer.
    (d) Use the Bayesian CLT to approximate the posterior of λ under the setting of (c), plot the approximate posterior, and compare the plot with the plot from (c).

4. Consider the model $Y_i|\sigma_i^2 \overset{indep}{\sim} \text{Normal}(0, \sigma_i^2)$ for $i = 1, ..., n$ where $\sigma_i^2|b \sim \text{InvGamma}(a, b)$ and $b \sim \text{Gamma}(1, 1)$.

    (a) Derive the full conditional posterior distributions for σ_1^2 and b.
    (b) Write pseudocode for Gibbs sampling, i.e., describe in detail each step of the Gibbs sampling algorithm.
    (c) Write your own Gibbs sampling code (not in JAGS) and plot the marginal posterior density for each parameter. Assume $n = 10$, $a = 10$ and $Y_i = i$ for $i = 1, ..., 10$.
    (d) Repeat the analysis with $a = 1$ and comment on the convergence of the MCMC chain.

(e) Implement this model in (c) using JAGS and compare the results with the results in (c).

5. Consider the model $Y_i|\mu, \sigma^2 \sim \text{Normal}(\mu, \sigma^2)$ for $i = 1, ..., n$ and $Y_i|\mu, \delta, \sigma^2 \sim \text{Normal}(\mu + \delta, \sigma^2)$ for $i = n+1, ..., n+m$, where $\mu, \delta \sim \text{Normal}(0, 100^2)$ and $\sigma^2 \sim \text{InvGamma}(0.01, 0.01)$.

(a) Give an example of a real experiment for which this would be an appropriate model.

(b) Derive the full conditional posterior distributions for μ, δ, and σ^2.

(c) Simulate a dataset from this model with $n = m = 50$, $\mu = 10$, $\delta = 1$, and $\sigma = 2$. Write your own Gibbs sampling code (not in JAGS) to fit the model above to the simulated data and plot the marginal posterior density for each parameter. Are you able to recover the true values reasonably well?

(d) Implement this model using JAGS and compare the results with the results in (c).

6. Fit the following model to the NBA free throw data in the table in the Exercise 17 in Chapter 1:

$$Y_i|\theta_i \sim \text{Binomial}(n_i, \theta_i) \text{ and } \theta_i|m \sim \text{Beta}[\exp(m)q_i, \exp(m)(1-q_i)],$$

where Y_i is the number of made clutch shots for player $i = 1, ..., 10$, n_i is the number of attempted clutch shots, $q_i \in (0, 1)$ is the overall proportion, and $m \sim \text{Normal}(0, 10)$.

(a) Explain why this is a reasonable prior for θ_i.

(b) Explain the role of m in the prior.

(c) Derive the full conditional posterior for θ_1.

(d) Write your own MCMC algorithm to compute a table of posterior means and 95% credible intervals for all 11 model parameters $(\theta_1, ..., \theta_{10}, m)$. Turn in commented code.

(e) Fit the same model in JAGS. Turn in commented code, and comment on whether the two algorithms returned the same results.

(f) What are the advantages and disadvantages of writing your own code as opposed to using JAGS in this problem and in general?

7. Open and plot the galaxies data in R using the code below,

```
> library(MASS)
> data(galaxies)
> ?galaxies
> Y <- galaxies
> n <- length(Y)
> hist(Y,breaks=25)
```

Model the observations $Y_1, ..., Y_{82}$ using the Student-t distribution with location μ, scale σ and degrees of freedom k. Assume prior distributions $\mu \sim \text{Normal}(0, 10000^2)$, $1/\sigma^2 = \tau \sim \text{Gamma}(0.01, 0.01)$ and $k \sim \text{Uniform}(1, 30)$.

(a) Give reasonable initial values for each of the three parameters.

(b) Fit the model using JAGS. Report trace plots of each parameter and discuss convergence.

(c) Graphically compare the t distribution with parameters set to their posterior mean with the observed data. Does the model fit the data well?

8. Download the galaxy data

```
> library(MASS)
> data(galaxies)
> Y <- galaxies
```

Suppose $Y_i | \boldsymbol{\theta} \overset{iid}{\sim} \text{Laplace}(\mu, \sigma)$ for $i = 1, \ldots, n$ where $\boldsymbol{\theta} = (\mu, \sigma)$.

(a) Assuming the improper prior $\sigma \sim \text{Uniform}(0, 100000)$ and $\pi(\mu) = 1$ for all $\mu \in (-\infty, \infty)$, plot the joint posterior distribution of $\boldsymbol{\theta}$ and the marginal posterior distributions of μ and σ.

(b) Compute the posterior mean of $\boldsymbol{\theta}$ from your analysis in (a) and plot the Laplace PDF with these values against the observed data. Does the model fit the data well?

(c) Plot the posterior predictive distribution (PPD) for a new observation $Y^* | \boldsymbol{\theta} \sim \text{Laplace}(\mu, \sigma)$. How do the mean and variance of the PPD compare to the mean and variance of the "plug-in" distribution from (b)?

9. In Section 2.4 we compared Reggie Jackson's home run rate in the regular season and World Series. He hit 563 home runs in 2820 regular-season games and 10 home runs in 27 World Series games (a player can hit 0, 1, 2, ... home runs in a game). Assuming Uniform(0,10) priors for both home run rates, use JAGS to summarize the posterior distribution of (i) his home run rate in the regular season, (ii) his home run rate in the World Series, and (iii) the ratio of these rates. Provide trace plots for all three parameters and discuss convergence of the MCMC sampler including appropriate convergence diagnostics.

10. As discussed in Section 1.6, [24] report that the number of marine bivalve species discovered each year from 2010-2015 was 64, 13, 33, 18, 30 and 20. Denote Y_t as the number of species discovered in year

2009 + t (so that $Y_1 = 64$ is the count for 2010). Use JAGS to fit the model

$$Y_t | \alpha, \beta \overset{indep}{\sim} \text{Poisson}(\lambda_t)$$

where $\lambda_t = \exp(\alpha + \beta t)$ and $\alpha, \beta \overset{indep}{\sim} \text{Normal}(0, 10^2)$. Summarize the posterior of α and β and verify that the MCMC sampler has converged. Does this analysis provide evidence that the rate of discovery is changing over time?

11. Write your own Metropolis sampler (i.e., not JAGS) for the previous problem. Turn in commented code, report your candidate distribution for each parameters and corresponding acceptance ratio, and use trace plots for each parameter to show the chains have converged.

12. A clinical trial assigned 100 patients each to placebo, the low dose of a new drug, and the high dose of the new drug (for a total of 300 patients); the data are given in the table below.

Treatment	Positive outcome	Negative outcome
Placebo	52	48
Low dose	60	40
High dose	54	46

Conduct a Bayesian analysis using JAGS with uniform priors on the probability of a patient having a positive outcome under each treatment.

(a) Report the posterior mean and 95% interval for the probability of a positive outcome for each treatment group.

(b) Compute the posterior probability that the low dose is the best of the three treatment options.

13. Consider the normal mixture model

$$Y_i | \theta \overset{iid}{\sim} f(y|\theta) = \frac{1}{2} [\phi(y - \theta) + \phi(y)],$$

for $i = 1, ..., n$ where $\theta \in \mathbb{R}$ and $\phi(z) = \exp\{-z^2/2\}/\sqrt{2\pi}$ denotes the density of a standard normal distribution. We have shown (Section 2.4, problem 14) that an improper prior cannot be used for this likelihood. In this problem we explore the use of vague but proper priors. Analyze data simulated in R as:

```
> set.seed(27695)
> theta_true <- 4
> n          <- 30
> B          <- rbinom(n,1,0.5)
> Y          <- rnorm(n,B*theta_true,1)
```

(a) Argue that the R code above generates samples from $f(y|\theta)$.

(b) Plot the simulated data, and plot $f(y|\theta)$ for $y \in [-3, 10]$ separately for $\theta = \{2, 4, 6\}$.

(c) Assuming prior $\theta \sim \text{Normal}(0, 10^2)$, obtain the MAP estimate of θ and its asymptotic posterior standard deviation using the following R code:

```
> library(stats4)
> nlp  <- function(theta,Y){
>    like           <- 0.5*dnorm(Y,0,1) +
>                       0.5*dnorm(Y,theta,1)
>    prior          <- dnorm(theta,0,10)
>    neg_log_post <- -sum(log(like)) - log(prior)
> return(neg_log_post)}
>
> map_est <- mle(nlp,start=list(theta=1),
>                         fixed=list(Y=Y))
> sd      <- sqrt(vcov(map_est))
```

(d) Suppose the prior distribution is $\theta \sim N(0, 10^k)$. Plot the posterior density of θ for $k \in \{0, 1, 2, 3\}$ and compare these posterior densities with the asymptotic normal distribution in (c) (by overlaying all five densities on the same plot).

(e) Use JAGS to fit this model via its mixture representation $Y_i|B_i, \theta \sim \text{Normal}(B_i\theta, 1)$, where $B_i \overset{iid}{\sim} \text{Bernoulli}(0.5)$ and $\theta \sim \text{Normal}(0, 10^2)$. Compare the posterior distribution of θ with the results from part (d).

14. Let $Y_1, \ldots, Y_n$ be a random sample from a shifted exponential distribution with density

$$f(y|\alpha, \beta) = \begin{cases} \beta \exp^{-\beta(y-\alpha)} & y \geq \alpha \\ 0 & y < \alpha \end{cases}$$

where $\alpha > 0$ and $\beta > 0$ are the parameters of interest. Assume prior distributions $\alpha \sim \text{Uniform}(0, c)$ and $\beta \sim \text{Gamma}(a, b)$.

(a) Plot $f(y|\alpha, \beta)$ for $\alpha = 2$ and $\beta = 3$ and $y \in [0, 5]$, give interpretations of α and β, and describe a real experiment where this model might be appropriate.

(b) Give the full conditional distributions of α and β and determine if they are members of a common family of distributions.

(c) Write pseudo code for an MCMC sampler including initial values, full details of each sampling step, and how you plan to assess convergence.

15. Using the data (`mass` and `age`) provided in Listing 3.7, fit the following non-linear regression model:

$$\texttt{mass}_i \sim \text{Normal}\left(\mu_i, \sigma^2\right) \text{ where } \mu_i = \theta_1 + \theta_2 \texttt{age}_i^{\theta_3},$$

with priors $\theta_1 \sim \text{Normal}(0, 100^2)$, $\theta_2 \sim \text{Uniform}(0, 20000)$, $\theta_3 \sim \text{Normal}(0, 1)$, and $\sigma^2 \sim \text{InvGamma}(0.01, 0.01)$.

(a) Fit the model in JAGS and plot the data (`age` versus `mass`) along with the posterior mean of μ_i to verify the model fits reasonably well.

(b) Conduct a thorough investigation of convergence.

(c) Give three steps you might take to improve convergence.

16. Assume the (overly complicated) model $Y|n, p \sim \text{Binomial}(n, p)$ with prior distributions $n \sim \text{Poisson}(\lambda)$ and $p \sim \text{Beta}(a, b)$. The observed data is $Y = 10$.

(a) Describe why convergence may be slow for this model.

(b) Fit the model in JAGS with $\lambda = 10$ and $a = b = 1$. Check convergence and summarize the posterior of n, p and $\theta = np$.

(c) Repeat the analysis in (b) except with $a = b = 10$. Comment on the effect of the prior distribution of p on convergence for all the three parameters.

17. Consider the (overly complicated) model

$$Y_i|\mu_i, \sigma_i^2 \sim \text{Normal}(\mu_i, \sigma_i^2)$$

where $\mu_i \overset{iid}{\sim} \text{Normal}(0, \theta_1^{-1})$, $\sigma_i^2 \sim \text{InvGamma}(\theta_2, \theta_3)$ and $\theta_j \sim \text{Gamma}(\epsilon, \epsilon)$ for $j = 1, 2, 3$. Assume the data are $Y_i = i$ for $i = 1, ..., n$.

(a) Describe why convergence may be slow for this model.

(b) Fit the model in JAGS with all four combinations of $n \in \{5, 25\}$ and $\epsilon \in \{0.1, 10\}$ and report the effective sample size for θ_1, θ_2, and θ_3 for all four fits.

(c) Comment on the effect of n and ϵ on convergence.

4

Linear models

CONTENTS

Linear models form the foundation for much of statistical modeling and intuition. In this chapter, we introduce many common statistical models and implement them in the Bayesian framework. We focus primarily on Bayesian aspects of these analyses including selecting priors, computation and comparisons with classical methods. The chapter begins with analyses of the mean of a normal population (Section 4.1.1) and comparison of the means of two normal populations (Section 4.1.2), which are analogous to the classic one-sample and two-sample t-tests. Section 4.2 introduces the more general Bayesian multiple linear regression model including priors that are appropriate for high-dimensional problems. Multiple regression is extended to non-Gaussian data in Section 4.3 via generalized linear models and correlated data in Section 4.4 via linear mixed models.

The Bayesian linear regression model makes strong assumptions including that the mean is a linear combination of the covariates, and that the observations are Gaussian and independent. Estimates are robust to small departures from these assumptions, but nonetheless these assumptions should be carefully evaluated (Section 5.6). Section 4.5 concludes with several extensions to the basic linear regression model to illustrate the flexibility of Bayesian modeling.

4.1   Analysis of normal means

4.1.1   One-sample/paired analysis

In a one-sample study, the n observations are modeled as $Y_i|\mu, \sigma^2 \overset{iid}{\sim}$ Normal(μ, σ^2) and the objective is to determine if $\mu = 0$. This model is often applied to experiments where each unit i actually has two measurements taken under different conditions and Y_i is the difference between the measurements. For example, say student i's math scores before and after a tutorial session are Z_{0i} and Z_{1i}, respectively, then testing if the (population) mean of $Y_i = Z_{1i} - Z_{0i}$ is zero is a way to evaluate the effectiveness of the tutorial sessions.

A Bayesian analysis specifies priors for μ and σ^2 and then summarizes the marginal posterior of μ accounting for uncertainty in σ^2. In this chapter we use the Jeffreys' prior, but conjugate normal/inverse gamma priors can also be used as in Section 3.2. In most cases it is sufficient to plot the posterior density $p(\mu|\mathbf{Y})$ and report the posterior mean and 95% interval. We will also address the problem using a formal hypothesis test. In this chapter we will compute the posterior probabilities of the one-sided hypotheses $H_0 : \mu \leq 0$ and $H_1 : \mu > 0$; in Chapter 5 we test the point hypothesis $H_0 : \mu = 0$ versus $H_1 : \mu \neq 0$.

Conditional distribution with the variance fixed: Conditional on σ^2, μ has Jeffreys prior $\pi(\mu) \propto 1$. The posterior is $\mu|\mathbf{Y} \sim$ Normal $\left(\bar{Y}, \frac{\sigma^2}{n} \right)$ and thus the $100(1 - \alpha)\%$ posterior credible interval for μ is

$$\bar{Y} \pm z_{\alpha/2} \frac{\sigma}{\sqrt{n}}, \tag{4.1}$$

where z_τ is the τ quantile of the standard normal distribution (i.e., the normal distribution with mean zero and variance one). This exactly matches the classic $100(1 - \alpha)\%$ confidence interval. While the credible interval and confidence interval are numerically identical in this case, they are interpreted differently. The Bayesian credible interval quantifies uncertainty about μ given this dataset, $\mathbf{Y}$, whereas the confidence interval quantifies the anticipated variation in $\bar{Y}$ if we were to repeat the experiment.

For the test of null hypothesis $H_0 : \mu \leq 0$ versus the alternative hypothesis

$H_1 : \mu > 0$, the posterior probability of the null hypothesis is

$$\text{Prob}(H_0|\mathbf{Y}) = \text{Prob}(\mu < 0|\mathbf{Y}) = \Phi(-Z) \qquad (4.2)$$

where Φ is the standard normal cumulative distribution function and $Z = \sqrt{n}\bar{Y}/\sigma$ and thus matches exactly with a frequentist p-value. By definition, $\Phi(z_\tau) = \tau$, and so the decision rule to reject H_0 in favor of H_1 if the posterior probability of H_0 is less than α is equivalent to rejecting H_0 if $-Z < z_\alpha$, or equivalently if $Z > z_{1-\alpha}$ ($-z_\alpha = z_{1-\alpha}$ due to the symmetry of the standard normal PDF). Therefore, the decision rule to reject H_0 in favor of H_1 at significance level α if $Z > z_{1-\alpha}$ is identical to the classic one-sided z-test. However, unlike the classical test, we can quantify our uncertainty using the posterior probability that the hypothesis H_0 (or H_1) is true since we have computed $\text{Prob}(H_0|\mathbf{Y})$.

Unknown variance: As shown in Section 2.3, the Jeffreys' prior for (μ, σ^2) is

$$\pi(\mu, \sigma^2) \propto \left(\frac{1}{\sigma^2}\right)^{3/2}. \qquad (4.3)$$

Appendix A.3 shows that the marginal posterior of μ integrating over σ^2 is

$$\mu|\mathbf{Y} \sim t_n \left(\bar{Y}, \hat{\sigma}^2/n\right), \qquad (4.4)$$

where $\hat{\sigma}^2 = \sum_{i=1}^n (Y_i - \bar{Y})^2/n$, i.e., the posterior is Student's t distribution with location $\bar{Y}$, variance parameter $\hat{\sigma}^2/n$ and n degrees of freedom. Posterior inference such as credible sets or the posterior probability that μ is positive follow from the quantiles of Student's t distribution. The credible set is slightly different than the frequentist confidence interval because the degrees of freedom in the classic t-test is $n-1$, whereas the degrees of freedom in the posterior is n; this is the effect of the prior on σ^2.

In classical statistics, when σ^2 is unknown the Z-test based on the normal distribution is replaced with the t-test based on Student's t distribution. Similarly, the posterior distribution of the mean changes from a normal distribution given σ^2 to Student's t distribution when uncertainty about the variance is considered. Figure 4.1 compares the Gaussian and t density functions. The density functions are virtually identical for $n = 25$, but for $n = 5$ the t distribution has heavier tails than the Gaussian distribution; this is the effect of accounting for uncertainty in σ^2.

4.1.2   Comparison of two normal means

The two-sample test compares the mean in two groups. For example, an experiment might take the blood pressure of a random sample of n_1 mail carriers that walk their route and n_2 mail carriers that drive their route to determine if these groups have different mean blood pressure. Let $Y_i \overset{iid}{\sim} \text{Normal}(\mu, \sigma^2)$ for $i = 1, ..., n_1$ and $Y_i \overset{iid}{\sim} \text{Normal}(\mu + \delta, \sigma^2)$ for $i = n_1 + 1, ..., n_1 + n_2 = n$, so that

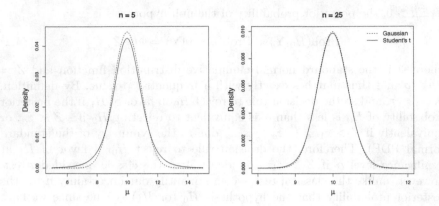

FIGURE 4.1
Comparison of the Gaussian and Student's t distributions. Below
are the Gaussian PDF with mean $\bar{Y}$ and standard deviation $\sigma/\sqrt{n}$ compared
to the PDF of Student's t distribution with location $\bar{Y}$, scale $\hat{\sigma}/\sqrt{n}$, and n
degrees of freedom. The plots assume $\bar{Y} = 10$, $\sigma = \hat{\sigma} = 2$, and $n \in \{5, 25\}$.

δ is the difference in means and the parameter of interest. Denote the sample
mean of the n_j observations in group $j = 1, 2$ as $\bar{Y}_j$ and the group-specific vari-
ance estimators as $s_1^2 = \sum_{i=1}^{n_1}(Y_i - \bar{Y}_1)^2/n_1$ and $s_2^2 = \sum_{i=n_1+1}^{n_1+n_2}(Y_i - \bar{Y}_2)^2/n_2$.

Conditional distribution with the variance fixed: Conditional on
the variance and flat prior $\pi(\mu, \delta) \propto 1$ it can be shown that the posterior of
the difference in means is

$$\delta|\mathbf{Y} \sim \text{Normal}\left(\bar{Y}_2 - \bar{Y}_1, \sigma^2\left[\frac{1}{n_1} + \frac{1}{n_2}\right]\right). \tag{4.5}$$

As in the one-sample case, posterior intervals and probabilities of hypotheses can be computed using the quantiles of the normal distribution. Also, as
in the one-sample case, the credible set and rejection rule match the classic confidence interval and one-sided z-test numerically but have a different
interpretations.

Unknown variance: The Jeffreys' prior for (μ, δ, σ^2) is (Section 2.3)

$$\pi(\mu, \delta, \sigma^2) \propto \frac{1}{(\sigma^2)^2}. \tag{4.6}$$

Appendix A.3 shows (as a special case of multiple linear regression, see Section
4.2) the marginal posterior distribution of δ integrating over both μ and σ^2 is

$$\delta|\mathbf{Y} \sim t_n\left[\bar{Y}_2 - \bar{Y}_1, \hat{\sigma}^2\left(\frac{1}{n_1} + \frac{1}{n_2}\right)\right], \tag{4.7}$$

Listing 4.1
R code for comparing two normal means with the Jeffreys' prior.

```
1   # Y1 is the n1-vector of data for group 1
2   # Y2 is the n2-vector of data for group 2
3
4   # Statistics from group 1
5   Ybar1 <- mean(Y1)
6   s21    <- mean((Y1-Ybar1)^2)
7   n1     <- length(Y1)
8
9   # Statistics from group 2
10  Ybar2 <- mean(Y2)
11  s22    <- mean((Y2-Ybar2)^2)
12  n2     <- length(Y2)
13
14  # Posterior of the difference assuming equal variance
15  delta_hat <- Ybar2-Ybar1
16  s2         <- (n1*s21 + n2*s22)/(n1+n2)
17  scale      <- sqrt(s2)*sqrt(1/n1+1/n2)
18  df         <- n1+n2
19  cred_int   <- delta_hat + scale*qt(c(0.025,0.975),df=df)
20
21  # Posterior of delta assuming unequal variance using MC sampling
22  mu1        <- Ybar1 + sqrt(s21/n1)*rt(1000000,df=n1)
23  mu2        <- Ybar2 + sqrt(s22/n2)*rt(1000000,df=n2)
24  delta      <- mu2-mu1
25
26  hist(delta,main="Posterior distribution of the difference in
          means")
27  quantile(delta,c(0.025,0.975)) # 95% credible set
```

where $\hat{\sigma}^2 = (n_1\hat{s}_1^2 + n_2\hat{s}_2^2)/n$ is a pooled variance estimator. As with the one-sample model, the difference between the posterior for the known-variance versus unknown-variance cases is that an estimate of σ^2 is inserted in the posterior and the Gaussian distribution is replaced with Student's t distribution with n degrees of freedom. In the Bayesian analysis we did not "plug in" an estimate of σ^2, rather, by accounting for its uncertainty and marginalizing over μ and σ^2 the posterior for δ happens to have a natural estimator of σ^2 in δ's scale. Listing 4.1 implements this method.

Unequal variance: If the assumption that the variance is the same for both groups is violated, then the two-sample model can be extended as $Y_i \overset{iid}{\sim}$ Normal(μ_1, σ_1^2) for $i = 1, ..., n_1$ and $Y_i \overset{iid}{\sim}$ Normal(μ_2, σ_2^2) for $i = n_1+1, ..., n_1+n_2$. Since no parameters are shared across the two groups, we can apply the one-sample model separately for each group to obtain

$$\mu_j|\mathbf{Y} \overset{indep}{\sim} t_{n_j}(\bar{Y}_j, s_j^2/n_j) \tag{4.8}$$

for $j = 1, 2$, and the posterior of the difference in means $\delta = \mu_2 - \mu_1$ follows. The posterior of δ is the difference between two Student t random variables, which does not in general have a simple form. However, the posterior can be approximated with arbitrary precision using Monte Carlo sampling as in Listing 4.1.

4.2 Linear regression

The multiple linear regression model with response (also called the dependent variable or outcome) Y_i and covariates (also called independent variables, predictors or inputs) $X_{i1}, ..., X_{ip}$ is

$$Y_i = \sum_{j=1}^{p} X_{ij}\beta_j + \varepsilon_i, \tag{4.9}$$

where $\boldsymbol{\beta} = (\beta_1, ..., \beta_p)^T$ are the regression coefficients and the errors (also called the residuals) are $\varepsilon_i \overset{iid}{\sim} \text{Normal}(0, \sigma^2)$. We assume throughout that $X_{i1} = 1$ for all i so that β_1 is the intercept (i.e., the mean if all other covariates are zero). This model includes as special cases the one-sample mean model in Section 4.1.1 (with $p = 1$ and $\beta_1 = \mu$) and the two-sample mean model in Section 4.1.2 (with $p = 2$, X_{i2} equal one if observation i is from the second group and zero otherwise, $\beta_1 = \mu$, and $\beta_2 = \delta$).

The coefficient β_j for $j > 0$ is the slope associated with the jth covariate. For the remainder of this subsection we will assume that all $p - 1$ covariates (excluding the intercept term) have been standardized to have mean zero and variance one so the prior can be specified without considering the scales of the covariates. That is, if the original covariate j, $\tilde{X}_{ij}$, had sample mean $\bar{X}_j$ and standard deviation $\hat{s}_j$ then we set $X_{ij} = (\tilde{X}_{ij} - \bar{X}_j)/\hat{s}_j$. After standardization, the slope β_j is interpreted as the change in the mean response corresponding to an increase of one standard deviation unit ($\hat{s}_j$) in the original covariate. Similarly, $\beta_j/\hat{s}_j$ is the expected increase in the mean response associated with an increase of one in the original covariate. The model actually has $p + 1$ parameters (p regression coefficients and variance σ^2) so we temporarily use p as the number of regression parameters and not the total number of parameters in the model.

The likelihood function for the linear model with n observations is the product of n Gaussian PDFs with means $\sum_{j=1}^{p} X_{ij}\beta_j$ and variance σ^2,

$$f(\mathbf{Y}|\boldsymbol{\beta}, \sigma^2) = \prod_{i=1}^{n} \frac{1}{\sqrt{2\pi}\sigma} \exp\left[-\frac{1}{2\sigma^2}\left(Y_i - \sum_{j=1}^{p} X_{ij}\beta_j\right)^2\right]. \tag{4.10}$$

Since maximizing the likelihood is equivalent to minimizing the negative log-likelihood, the maximum likelihood estimator for β can be written

$$\beta_{LS} = \arg\min_{\beta} \sum_{i=1}^{n} \left(Y_i - \sum_{j=1}^{p} X_{ij}\beta_j \right)^2. \tag{4.11}$$

Therefore, assuming Gaussian errors the maximum likelihood estimator is also the least squares estimator.

The model is conveniently written in matrix notation. Let $\mathbf{Y} = (Y_1, ..., Y_n)^T$ be the response vector of length n, and the design matrix $\mathbf{X}$ be the $n \times p$ matrix with the first column equal to the vector of ones for the intercept and column j having elements $X_{1j}, ..., X_{nj}$. The linear regression model is then

$$\mathbf{Y} \sim \text{Normal}(\mathbf{X}\beta, \sigma^2 \mathbf{I}_n) \tag{4.12}$$

where $\mathbf{I}_n$ is the $n \times n$ identity matrix with ones on the diagonal (so all responses have variance σ^2) and zeros off the diagonal (so the responses are uncorrelated). The usual least squares estimator in matrix notation is

$$\hat{\beta}_{LS} = (\mathbf{X}^T\mathbf{X})^{-1}\mathbf{X}^T\mathbf{Y}. \tag{4.13}$$

Note that the least squares estimator is unique only if $\mathbf{X}^T\mathbf{X}$ is full rank (i.e., $p < n$ and none of $\mathbf{X}$'s columns are redundant) and the estimator is poorly defined if $\mathbf{X}^T\mathbf{X}$ is not full rank. Assuming $\mathbf{X}$ is full rank, the sampling distribution used to construct frequentist confidence intervals and p-values is $\hat{\beta}_{LS} \sim \text{Normal}\left(\beta_0, \sigma^2(\mathbf{X}^T\mathbf{X})^{-1}\right)$, where β_0 is the true value.

4.2.1 Jeffreys prior

Conditional distribution with the variance fixed: Conditioned on σ^2, the Jeffreys prior is $\pi(\beta) \propto 1$. This improper prior only leads to a proper posterior if $\mathbf{X}^T\mathbf{X}$ has full rank, which is the same condition needed for the least squares estimator to be unique. Assuming this condition is met, the posterior is

$$\beta|\mathbf{Y}, \sigma^2 \sim \text{Normal}\left[\hat{\beta}_{LS}, \sigma^2(\mathbf{X}^T\mathbf{X})^{-1}\right]. \tag{4.14}$$

The posterior mean is the least squares solution and the posterior covariance matrix is the covariance matrix of the sampling distribution of the least squares estimator. Therefore, the posterior credible intervals from this model will numerically match the confidence intervals from a least squares analysis with known error variance.

Unknown variance: With σ^2 unknown, the Jeffreys' prior is $\pi(\beta, \sigma^2) \propto (\sigma^2)^{-p/2-1}$ (Section 2.3). Assuming $\mathbf{X}^T\mathbf{X}$ has full rank, Appendix A.3 shows that

$$\beta|\mathbf{Y} \sim t_n\left[\hat{\beta}_{LS}, \hat{\sigma}^2(\mathbf{X}^T\mathbf{X})^{-1}\right], \tag{4.15}$$

Listing 4.2
R code for Bayesian linear regression under the Jeffreys' prior.

```
1   # This code assumes:
2   #   Y is the n-vector of observations
3   #   X us the n x p matrix of covariates
4   #   The first column of X is all ones for the intercept
5
6   # Compute posterior mean and 95% interval
7   beta_mean   <- solve(t(X)%*%X)%*%t(X)%*%Y
8   sigma2      <- mean((Y-X%*%beta_mean)^2)
9   beta_cov    <- sigma2*solve(t(X)%*%X)
10  beta_scale  <- sqrt(diag(beta_cov))
11  df          <- length(Y)
12  beta_025    <- beta_mean + beta_scale*qt(0.025,df=df)
13  beta_975    <- beta_mean + beta_scale*qt(0.975,df=df)
14
15  # Package the output
16  out            <- cbind(beta_mean,beta_025,beta_975)
17  rownames(out) <- colnames(X)
18  colnames(out) <- c("Mean","Q 0.025","Q 0.975")
```

where $\hat{\sigma}^2 = (\mathbf{Y} - \mathbf{X}\hat{\boldsymbol{\beta}}_{LS})^T(\mathbf{Y} - \mathbf{X}\hat{\boldsymbol{\beta}}_{LS})/n$. That is, the marginal posterior of $\boldsymbol{\beta}$ follows the p-dimensional Student's t distribution with location $\hat{\boldsymbol{\beta}}_{LS}$, covariance matrix $\hat{\sigma}^2(\mathbf{X}^T\mathbf{X})^{-1}$, and n degrees of freedom.

A property of the multivariate t distribution is that the marginal distribution of each element is univariate t so that

$$\beta_j|\mathbf{Y} \sim t_n(\hat{\beta}_j, s_j^2) \tag{4.16}$$

where $\hat{\beta}_j$ is the jth element of $\hat{\boldsymbol{\beta}}_{LS}$ and s_j^2 is the jth diagonal element of $\hat{\sigma}^2(\mathbf{X}^T\mathbf{X})^{-1}$. Therefore, both the joint distribution of all the regression coefficients and marginal distribution of each regression coefficient belong to a known family of distributions, and can thus be computed without MCMC sampling. Listing 4.2 provides R code to compute the posterior mean and 95% intervals.

4.2.2 Gaussian prior

For high-dimensional cases with more predictors than observations the improper Jeffreys' prior does not lead to a valid posterior distribution and thus a proper prior is required. Even in less extreme cases with moderate p, a proper prior can stabilize the posterior and give better results than the improper prior. The conjugate prior (conditioned on σ^2) for $\boldsymbol{\beta}$ is $\boldsymbol{\beta}|\sigma^2 \sim \text{Normal}(\boldsymbol{\mu}, \sigma^2\boldsymbol{\Omega})$. We include σ^2 in the prior variance to account for the scale of the response. Typically the prior is centered around zero,

Listing 4.3
JAGS code for multiple linear regression with Gaussian priors.

```
1   # Likelihood
2   for(i in 1:n){
3      Y[i] ~ dnorm(inprod(X[i,],beta[]),taue)
4   }
5   # Priors
6   beta[1] ~ dnorm(0,0.001) #X[i,1]=1 for the intercept
7   for(j in 2:p){
8      beta[j] ~ dnorm(0,taub*taue)
9   }
10  taue ~ dgamma(0.1, 0.1)
11  taub ~ dgamma(0.1, 0.1)
```

and so from here on we set $\mu = 0$. This prior combined with the likelihood $\mathbf{Y} \sim \text{Normal}(\mathbf{X}\beta, \sigma^2 \mathbf{I}_n)$ gives posterior

$$\beta | \mathbf{Y}, \sigma^2 \sim \text{Normal} \left[(\mathbf{X}^T\mathbf{X} + \mathbf{\Omega}^{-1})^{-1}\mathbf{X}^T\mathbf{Y}, \sigma^2(\mathbf{X}^T\mathbf{X} + \mathbf{\Omega}^{-1})^{-1} \right], \quad (4.17)$$

which is a proper distribution as long as the prior is proper ($\mathbf{\Omega}$ is positive definite) even if $p > n$ or the covariates are perfectly collinear.

There are several choices for the prior covariance matrix $\mathbf{\Omega}$. The most common is the diagonal matrix $\mathbf{\Omega} = \tau^2 \mathbf{I}_p$, which is equivalent to the prior $\beta_j | \sigma^2 \overset{iid}{\sim} \text{Normal}(0, \sigma^2\tau^2)$. Under this prior the MAP estimator is the ridge regression estimator [44]

$$\beta_R = \arg\min_{\beta} \sum_{i=1}^{n} \left(Y_i - \sum_{j=1}^{p} X_{ij}\beta_j \right)^2 + \lambda \sum_{j=1}^{p} \beta_j^2, \quad (4.18)$$

where $\lambda = 1/\tau^2$. Ridge regression is often used to stabilize least squares problems when the number of predictors is large and/or the covariates are collinear. In ridge regression, the tuning parameter λ can be selected based on cross-validation. In a fully Bayesian analysis, τ^2 can either be fixed to a large value to give an uninformative prior, or given a conjugate inverse gamma prior as in Listing 4.3 to allow the data to determine how much to shrink the coefficients towards zero. If τ^2 is given a prior, then the intercept term β_1 should be given a different variance because it plays a different role than the other regression coefficients.

The form of the posterior simplifies under Zellner's g-prior [83] $\beta | \sigma^2 \sim \text{Normal} \left[0, \frac{\sigma^2}{g}(\mathbf{X}^T\mathbf{X})^{-1} \right]$ for $g > 0$. The conditional posterior is then

$$\beta | \mathbf{Y}, \sigma^2 \sim \text{Normal} \left[c\hat{\beta}_{LS}, c\sigma^2(\mathbf{X}^T\mathbf{X})^{-1} \right], \quad (4.19)$$

where $c = 1/(g+1) \in (0,1)$ is the shrinkage factor. This speeds computation because $\hat{\boldsymbol{\beta}}_{LS}$ and $(\mathbf{X}^T\mathbf{X})^{-1}$ can be computed once outside the MCMC sampler. The shrinkage factor c determines how strongly the posterior mean and covariance are shrunk towards zero. A common choice is $g = 1/n$ and thus $c = n/(n+1)$. Since Fisher's information matrix for the Gaussian distribution is the inverse covariance matrix, the prior contributes $1/n^{th}$ the information as the likelihood, and so this prior is called the unit information prior [49].

4.2.3   Continuous shrinkage priors

In regressions with many predictors it is often assumed that most of the p predictors have little effect on the response. The Gaussian prior can reflect this prior belief with a small prior variance, but a small prior variance has the negative side effect of shrinking the posterior mean of the important regression coefficients toward zero and thus inducing bias. An alternative is to use double exponential priors for the β_j that have (Figure 4.2) a peak at zero to reflect the prior belief that most predictors have no effect on the response, but a heavy tail to reflect the prior belief that there are a few predictors with strong effects. Listing 4.4 provides JAGS code to implement this model (although the R package BLR [21] is likely faster).

Assuming the model $Y_i \sim \text{Normal}(\sum_{j=1}^{p} X_{ij}\beta_j, \sigma^2)$ with double exponential priors $\beta_j \stackrel{iid}{\sim} \text{DE}(\lambda/\sigma^2)$ and thus prior PDF $\pi(\beta_j) \propto \exp\left(-\frac{\lambda}{2\sigma^2}|\beta_j|\right)$, then the maximum a posterior (MAP) estimate is (since maximizing the posterior is equivalent to minimizing twice the negative log posterior)

$$\hat{\boldsymbol{\beta}}_{LASSO} = \arg\min_{\boldsymbol{\beta}} \sum_{i=1}^{n} \left(Y_i - \sum_{j=1}^{p} X_{ij}\beta_j \right)^2 + \lambda \sum_{j=1}^{p} |\beta_j|. \qquad (4.20)$$

This is the famous LASSO [81] penalized regression estimator and thus the double exponential prior is often called the Bayesian LASSO prior. An attractive feature of this estimator is that some of the estimates may have $\hat{\beta}_j$ set exactly to zero, and this then performs variable selection simultaneously with estimation. In other words, the LASSO encodes the prior belief that some of the covariates are unimportant.

The double exponential prior is just one example of a shrinkage prior with peak at zero and heavy tails. The horseshoe prior [16] is $\beta_j|\lambda_j \sim$ Normal$(0, \lambda_0^2\lambda_j^2)$ where λ_0 is global variance common to all regression coefficients and λ_j is a local prior standard deviation specific to β_j. The local variances are given half-Cauchy prior (i.e., student-t prior with one degree of freedom restricted to be positive). This global-local prior is designed to shrink null coefficients towards zero by having small variance while the true signals have uninformative priors with large variance. The Dirichlet–Laplace prior [11] gives even more shrinkage towards zero by supplementing the Bayesian LASSO with local shrinkage parameters and a Dirichlet prior on the shrinkage parameters. The R2D2 prior [84] is another global-local shrinkage prior for

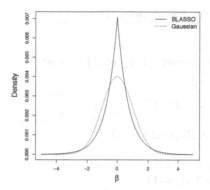

FIGURE 4.2
Comparison of the Gaussian and double exponential prior distributions. Below are the standard normal PDF and the double exponential PDF with parameters set to give mean zero and variance one.

the regression coefficients that is constructed so that the model's coefficient of determination (i.e., R-squared) has a beta prior. The R2D2 prior has the most mass at zero and heaviest tail among these priors. For an alternative approach that places positive prior probability on the regression coefficients being exactly zero see the spike-and-slab prior in Section 5.3.

4.2.4 Predictions

One use of linear regression is to make a prediction for a new set of covariates, $\mathbf{X}^{pred} = (X_1^{pred}, ..., X_p^{pred})$. Given the model parameters, the distribution of

Listing 4.4
JAGS code for the Bayesian LASSO.

```
1   # Likelihood
2   for(i in 1:n){
3       Y[i] ~ dnorm(inprod(X[i,],beta[]),taue)
4   }
5   # Priors
6   beta[1] ~ dnorm(0,0.001)
7   for(j in 2:p){
8       beta[j] ~ ddexp(0,taub*taue)
9   }
10  taue  ~  dgamma(0.1, 0.1)
11  taub  ~  dgamma(0.1, 0.1)
```

Listing 4.5
JAGS code for linear regression predictions.

```
1    # Likelihood
2    for(i in 1:n){
3        Y[i] ~ dnorm(inprod(X[i,],beta[]),taue)
4    }
5    # Priors
6    beta[1] ~ dnorm(0,0.001)
7    for(j in 2:p){
8        beta[j] ~ dnorm(0,taub*taue)
9    }
10   taue  ~ dgamma(0.1, 0.1)
11   taub  ~ dgamma(0.1, 0.1)
12
13   # Predictions
14   for(i in 1:n_pred){
15       Y_pred[i] ~ dnorm(inprod(X_pred[i,],beta[]),taue)
16   }
17   # User must pass JAGS the covariates X_pred and integer n_pred
18   # JAGS returns PPD samples of Y_pred
```

the new response is $Y^{pred}|\boldsymbol{\beta},\sigma^2 \sim \text{Normal}(\sum_{j=1}^{p} X_j^{pred}\beta_j,\sigma^2)$. To properly account for parametric uncertainty, we should use the posterior predictive distribution (Section 1.5) that averages over the uncertainty in $\boldsymbol{\beta}$ and σ^2. MCMC provides a means to sample from the PPD by making a sample from the predictive distribution for each of the $s = 1,...,S$ MCMC samples of the parameters, $Y^{(s)}|\boldsymbol{\beta}^{(s)},\sigma^{2(s)} \sim \text{Normal}(\sum_{j=1}^{p} X_j^{pred}\beta_j^{(s)},\sigma^{2(s)})$, and using the S draws $Y^{(1)},...,Y^{(S)}$ to approximate the PPD. The PPD is then summarized the same way as other posterior distributions, such as by the posterior mean and 95% interval. Similar approaches can use to analyze missing data as in Section 6.4.

Listing 4.5 gives JAGS code to make linear regression predictions. The matrix of predictors X_{pred} must be passed to JAGS and JAGS will return the predictions Y_{pred}. Making predictions with JAGS can slow the sampler and consume memory, and so it is often better to first perform MCMC sampling for the parameters using JAGS and then make predictions in R as in Listing 4.6.

4.2.5   Example: Factors that affect a home's microbiome

We use the data from [5], downloaded from http://figshare.com/articles/1000homes/1270900. The data are dust samples from the ledges above doorways from $n = 1,059$ homes (after removing samples with missing data; for missing data methods see Section 6.4) in the continental US. Bioinformatics processing detects the presence or absence of 763 species (technically

Listing 4.6

R code to use JAGS MCMC samples for linear regression predictions.

```
1   # INPUTS
2   #   beta_samples := S x p matrix of MCMC samples (from JAGS)
3   #   taue_samples := S x 1 matrix of MCMC samples (from JAGS)
4   #   X_pred       := n_pred x p matrix of prediction covariates
5
6   S       <- nrow(beta_samples)
7   n_pred <- nrow(X_pred)
8   Y_pred <- matrix(NA,S,n_pred)
9   sigma   <- 1/sqrt(taue_samples)
10
11  for(s in 1:S){
12    Y_pred[s,] ~ X_pred%*%beta_samples[s,]+rnorm(n_pred,0,sigma[s])
13  }
14
15  # OUTPUT
16  #   Y_pred       := S x n_pred matrix of PPD samples
```

operational taxonomic units) of fungi. The response is the log of the number of fungi species present in the sample, which is a measure of species richness. The objective is to determine which factors influence a home's species richness. For each home, eight covariates are included in this example: longitude, latitude, annual mean temperature, annual mean precipitation, net primary productivity (NPP), elevation, the binary indicator that the house is a single-family home, and the number of bedrooms in the home. These covariates are all centered and scaled to have mean zero and variance one.

We apply the Gaussian model in Listing 4.3 first with $\beta_j \overset{iid}{\sim}$ Normal$(0, 100^2)$ ("Flat prior") and then with $\beta_j | \sigma^2, \tau^2 \overset{iid}{\sim}$ Normal$(0, \sigma^2 \tau^2)$ with $\tau^2 \sim$ InvGamma$(0.1, 0.1)$ ("Gaussian shrinkage prior"), and the Bayesian LASSO prior in Listing 4.4. For each of the three models we ran two MCMC chains with 10,000 samples in the burn-in and 20,000 samples after burn-in. Trace plots (not shown) showed excellent convergence and the effective sample size exceeded 1,000 for all parameters and all models.

The results are fairly similar for the three priors (Figure 4.3). In all three models, temperature, NPP, elevation, and single-family home are the most important predictors, with the most richness estimated to occur in single-family homes with low temperature, NPP, and elevation. In all three models, the sample with largest fitted value (i.e., the posterior mean of $\mathbf{X}\beta$) is a single-family home with three bedrooms in Montpelier, VT, and the sample with the smallest fitted value is a multiple-family home with two bedrooms in Tempe, AZ.

Although the results are not that sensitive to the prior in this analysis, there are some notable difference. For example, compared to the posterior under a flat priors, the posterior of the slope for latitude (top right panel in

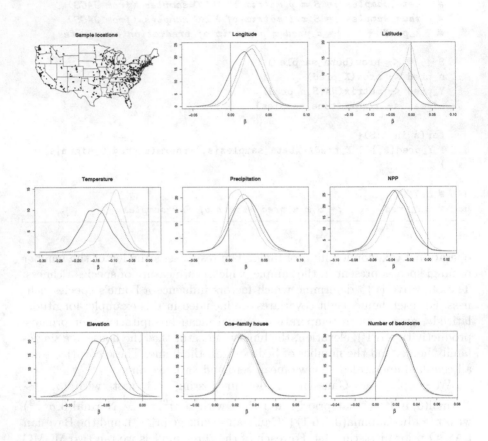

FIGURE 4.3
Regression analysis of the richness of a home's microbiome. The
first panel shows the sample locations and the remaining panels plot the posterior distributions of the regression coefficients, β_j. The three models are
distinguished by their priors for β_j: the flat prior is $\beta_j \sim \text{Normal}(0, 100^2)$
(solid line), the Gaussian shrinkage prior is $\beta_j \overset{iid}{\sim} \text{Normal}(0, \sigma^2\tau^2)$ with
$\tau^2 \sim \text{InvGamma}(0.1, 0.1)$ (dashed line), and the Bayesian LASSO is $\beta_j \overset{iid}{\sim}$
$\text{DE}(0, \sigma^2\tau^2)$ with $\tau^2 \sim \text{InvGamma}(0.1, 0.1)$ (dotted line).

Figure 4.3) is shrunk towards zero by the Gaussian shrinkage model, and the posterior density concentrates even more around the origin for the Bayesian LASSO prior. However, it is not clear from this plot which of these three fits in preferred; model comparison is discussed in Chapter 5.

4.3 Generalized linear models

Multiple linear regression assumes that the response variable is Gaussian and thus the mean response can be any real number. Many analyses do not conform to this assumption. For example, in Section 4.3.1 we analyze binary responses with support $\{0, 1\}$ and in Section 4.3.2 we analyze count data with support $\{0, 1, 2, ...\}$. Clearly these data are not Gaussian and their mean cannot be any real number because the mean must be between zero and one for binary data and positive for count data. The generalized linear model (GLM) extends linear regression concepts to handle these non-Gaussian outcomes. There is a deep theory of GLMs (exponential families, canonical links, etc; see [54]), but here we focus only on casting the GLM in the Bayesian framework through a few examples.

The basic steps to selecting a GLM are (1) determine the support of the response and select an appropriate parametric family and (2) link the covariates to the parameters that define this family. As an example, consider the Gaussian linear regression model in Section 4.2. If the support of the response is $(-\infty, \infty)$, a natural parametric family is the Gaussian distribution. Of course, there are other families with this support and the fit of the Gaussian family to the data should be verified empirically. Once the Gaussian family has been selected, the covariates must be linked to one or both of the parameters, the mean or the variance. Let the linear combination of the covariates be

$$\eta_i = \sum_{j=1}^{p} X_{ij}\beta_j. \tag{4.21}$$

The linear predictor η_i can take any value in $(-\infty, \infty)$ depending on X_{ij}. Therefore, to link the covariates with the mean we can simply set $\mathrm{E}(Y_i) = \eta_i$ as in standard linear regression. To complete the standard model we elect not to link the covariates with the variance, and simply set $\mathrm{V}(Y_i) = \sigma^2$ for all i.

The function that links the linear predictor with a parameter is called the link function. Say that the parameter in the likelihood for the response is θ_i (e.g., $\mathrm{E}(Y_i) = \theta_i$ or $\mathrm{V}(Y_i) = \theta_i$), then the link function g is

$$g(\theta_i) = \eta_i. \tag{4.22}$$

The link function must be an invertible function that is well-defined for all permissible values of the parameter. For example, in the Gaussian case the

Listing 4.7

Model statements for several GLMs in JAGS.

```
1
2    # (a) Logistic regression
3      for(i in 1:n){
4        Y[i]            ~ dbern(q[i])
5        logit(q[i]) <- inprod(X[i,],beta[])
6      }
7      for(j in 1:p){beta[j] ~ dnorm(0,0.01)}
8
9    # (b) Probit regression
10     for(i in 1:n){
11       Y[i]              ~ dbern(q[i])
12       probit(q[i]) <- inprod(X[i,],beta[])
13     }
14     for(j in 1:p){beta[j] ~ dnorm(0,0.01)}
15
16   # (c) Poisson regression
17     for(i in 1:n){
18       Y[i]                ~ dpois(lambda[i])
19       log(lambda[i]) <- inprod(X[i,],beta[])
20     }
21     for(j in 1:p){beta[j] ~ dnorm(0,0.01)}
22
23   # (d) Negative binomial regression
24     for(i in 1:n){
25       Y[i]              ~ dnegbin(q[i],m)
26       q[i]              <- m/(m + lambda[i])
27       log(lambda[i]) <- inprod(X[i,],beta[])
28     }
29     for(j in 1:p){beta[j] ~ dnorm(0,0.01)}
30     m ~ dgamma(0.1,0.1)
31
32   # (e) Zero-inflated Poisson
33     for(i in 1:n){
34       Y[i]                ~ dpois(q[i])
35       q[i]                <- Z[i]*lambda[i]
36       Z[i]                ~ dbern(p[i])
37       log(lambda[i]) <- inprod(X[i,],beta[])
38       logit(p[i])      <- inprod(X[i,],alpha[])
39     }
40     for(j in 1:p){beta[j] ~ dnorm(0,0.01)}
41     for(j in 1:p){alpha[j] ~ dnorm(0,0.01)}
42
43   # (f) Beta regression
44     for(i in 1:n){
45       Y[i]              ~ dbeta(r*q[i],r*(1-q[i]))
46       logit(q[i]) <- inprod(X[i,],beta[])
47     }
48     for(j in 1:p){beta[j] ~ dnorm(0,0.01)}
49     r ~ dgamma(0.1,0.1)
```

link function for the mean is the identity function $g(x) = x$ so that the mean can be any real number. To link the covariates to the variance we must ensure that the variance is positive, and so natural-log function $g(x) = \log(x)$ is more appropriate. Link functions are not unique, and must be selected by the user. For example, the link function for the mean could be replaced by $g(x) = x^3$ and the link function for the variance could be replaced with $g(x) = \log_{10}(x)$.

Bayesian fitting of a GLM requires selecting the prior and computing the posterior distribution. The priors for the regression coefficients discussed for Gaussian data in Section 4.2 can be applied for GLMs. The posterior distributions for GLMs are usually too complicated to derive the posterior in closed-form and prove that, say, a particular prior distribution leads to a student-t posterior distribution with interpretable mean and covariance. However, much of the intuition developed with Gaussian linear models carries over to GLMs.

With non-Gaussian responses the full conditional distributions for the β_j are usually not conjugate and Metropolis sampling must be used. Maximum likelihood estimates can be used as initial values and the corresponding standard errors can suggest appropriate candidate distributions. In the examples in this chapter we use JAGS to carry out the MCMC sampling. R also has dedicated packages for Bayesian GLMs, such as the MCMClogit package for logistic regression (Section 4.3.1) that are likely faster than JAGS. With flat priors and a large sample it is even more efficient to evoke the Bayesian central limit theorem (Section 3.1.3) and approximate the posterior as Gaussian (e.g., using the glm function in R) and avoid MCMC altogether.

4.3.1 Binary data

Binary outcomes $Y_i \in \{0, 1\}$ occur frequently when the result of an experiment is recorded only as an indicator of a success or failure. For example, in Section 1.2 the response was the binary indicator that a test for HIV was positive. Binary variables must follow a Bernoulli distribution. The Bernoulli distribution has one parameter, the success probability $\text{Prob}(Y_i = 1) = q_i \in [0, 1]$. Since the parameter is a probability, the link function must have input range $[0, 1]$ and output range $[-\infty, \infty]$. Below we discuss two such link functions: logistic and probit.

Logistic regression: The logistic link function is

$$g(q) = \text{logit}(q) = \log\left(\frac{q}{1 - q}\right). \tag{4.23}$$

This link function converts the event probability q first to the odds of the event, $q/(1 - q) > 0$, and then to the log odds, which can be any real number. The logistic regression model is written

$$Y_i \overset{indep}{\sim} \text{Bernoulli}(q_i) \text{ and } \text{logit}(q_i) = \eta_i = \sum_{j=1}^{J} X_{ij}\beta_j. \tag{4.24}$$

The inverse logistic function is $g^{-1}(x) = \exp(x)/[1+\exp(x)]$ and so the model can also be expressed as $Y_i \overset{indep}{\sim} \text{Bernoulli}(\exp(\eta_i)/[1+\exp(\eta_i)])$. JAGS code for this model is in Listing 4.7a.

Because the log odds of the event that $Y_i = 1$ are linear in the covariates, β_j is interpreted as the increase in the log odds corresponding to an increase of one in X_j with all other covariates held fixed. Similarly, with all over covariates held fixed, increasing X_j by one multiplies the odds by $\exp(\beta_j)$. Therefore if $\beta_j = 2.3$, increasing X_j by one multiplies the odds by ten and if $\beta_j = -2.3$, increasing X_j by one divides the odds by ten. This interpretation is convenient for communicating the results and specifying priors. For example, if a change of one in the covariate is deemed a large change, then a standard normal prior may have sufficient spread to represent an uninformative prior.

Probit regression: There are many possible link functions from $[0, 1]$ to $[-\infty, \infty]$. In fact, the quantile function (inverse CDF) of any continuous random variable with support $[-\infty, \infty]$ would suffice. The link function in logistic regression is the quantile function of the logistic distribution. In probit regression, the link function is the Gaussian quantile function

$$Y_i \overset{indep}{\sim} \text{Bernoulli}(q_i) \text{ and } q_i = \Phi(\eta_i), \tag{4.25}$$

where Φ is the standard normal CDF (Listing 4.7b).

Unfortunately, the regression coefficients β_j in probit regression do not have a nice interpretation such as their log-odds interpretation in logistic regression. However, probit regression is useful because it leads to Gibbs sampling and can be used to model dependence between binary variables. Probit regression is equivalent to specifying latent variables $Z_i \overset{indep}{\sim} \text{Normal}(\eta_i, \sigma^2)$ and assuming that observing $Y_i = 1$ indicates that Z_i exceeds the threshold z. For example, Z_i might represent a patient's blood pressure and Y_i is the corresponding binary indicator that the patient has high blood pressure (defined as exceeding z). In this example, the probability that the patient has high blood pressure is

$$
\begin{aligned}
\text{Prob}(Y_i = 1) &= \text{Prob}(Z_i > z) \\
&= \text{Prob}[(Z_i - \eta_i)/\sigma > (z - \eta_i)/\sigma] \\
&= 1 - \Phi[(z - \eta_i)/\sigma] \\
&= \Phi[(\eta_i - z)/\sigma] \\
&= \Phi\left[\left(\sum_{j=1}^{J} X_{ij}\beta_j - z\right)/\sigma\right].
\end{aligned}
$$

Since we never observe the latent variables Z_i we cannot estimate the threshold z or the variance σ^2. The threshold z is coupled with the intercept because adding a constant to the threshold and subtracting the same constant from the intercept does not affect the event probabilities q_i. Therefore, the threshold is typically set to $z = 0$. Similarly, multiplying σ and dividing the slopes β_j by the same constant does not affect the event probabilities, and so the variance is typically set to $\sigma^2 = 1$. This gives the usual probit regression model $q_i = \Phi(\eta_i)$.

In this formulation, the regression coefficients β have conjugate full conditional distributions conditioned on the latent Z_i. This leads to Gibbs sampling if the Z_i are imputed at each MCMC iteration [3]. Also, dependence between binary outcomes Y_i can be induced by a multivariate normal model for the latent variables Z_i.

4.3.2 Count data

Random variables with support $Y_i \in \{0, 1, 2, ...\}$ often arise as the number of events that occur in a time interval or spatial region. For example, in Section 2.1 we analyze the number of NFL concussions by season. In this chapter, we will focus on modeling the mean as a function of the covariates. The link between the linear predictor and the mean must ensure that $E(Y_i) = \lambda_i \geq 0$. A natural link function is the log link,

$$\log(\lambda_i) = \sum_{j=1}^{p} X_{ij}\beta_j. \tag{4.26}$$

In this model, the mean is multiplied by $\exp(\beta_j)$ if X_j increases by one with all other covariates remaining fixed. Unlike binary data, specifying the mean does not completely determine the likelihood for count data. We discuss below two families of distributions for the likelihood function: Poisson and negative binomial.

Poisson regression: The Poisson regression model is

$$Y_i|\lambda_i \overset{indep}{\sim} \text{Poisson}(\lambda_i) \text{ where } \lambda_i = \exp\left(\sum_{j=1}^{p} X_{ij}\beta_j\right). \tag{4.27}$$

As with logistic regression, the regression coefficients $\beta_1, ..., \beta_p$ can be given the priors discussed in Section 4.2, and Metropolis sampling can be used to explore the posterior (Listing 4.7c). A critical assumption of the Poisson model is that the mean and variance are equal, that is,

$$E(Y_i) = V(Y_i) = \lambda_i. \tag{4.28}$$

The distribution of a count is over-dispersed (under-dispersed) if its variance is greater than (less than) its mean. If over-dispersion is present, then the Poisson model is inappropriate and another model should be considered so that the posterior accurately reflects all sources of variability.

Negative binomial regression: One approach to accommodating over-dispersion is to incorporate gamma random variables $e_i \overset{iid}{\sim} \text{Gamma}(m, m)$

with mean 1 and variance $1/m$. Then

$$Y_i|e_i, \lambda_i, m \overset{indep}{\sim} \text{Poisson}(\lambda_i e_i). \tag{4.29}$$

Marginally over e_i, Y_i follows the negative binomial distribution,

$$Y_i|\lambda_i, m \sim \text{NegBinomial}(q_i, m) \tag{4.30}$$

with probability $q_i = m/(\lambda_i + m)$ and size m. The size m need not be an integer, but if it is and we envision a sequence of independent Bernoulli trials each with success probability q_i, then Y_i can be interpreted as the number of failures that occur before the mth success. More importantly for handling over-dispersion is that $\text{E}(Y_i) = \lambda_i$ and $\text{V}(Y_i) = \lambda_i + \lambda_i^2/m > \lambda_i$. The parameter m controls over-dispersion. If m is large, then $e_i \approx 1$ and the model reduces to the Poisson regression model with $\text{V}(Y_i) \approx \text{E}(Y_i)$; if m is close to zero, then e_i has large variance and thus $\text{V}(Y_i) > \text{E}(Y_i)$. The over-dispersion parameter can be given a gamma prior, as in Listing 4.7d.

Zero-inflated Poisson: Another common deviation from the Poisson distribution is an excess of zeros. For example, say that Y_i is the number of fish caught by visitor i to a state park. It may be that most of the Y_i are zero because only a small proportion (say p) of the visitors fished, but the distribution of Y_i for those that did fish is Poisson with mean λ. The probability of an observation being zero is then $1 - p$ for the non-fishers plus p times the Poisson probability at zero for the fishers. The PMF corresponding to this scenario is

$$f(y|p, \lambda) = \begin{cases} (1-p) + pf_P(0|\lambda) & \text{if } y = 0 \\ pf_P(y|\lambda) & \text{if } y > 0 \end{cases} \tag{4.31}$$

where f_P is the Poisson PMF. This is equivalent to the two-stage model

$$Y_i|Z_i, \lambda \sim \text{Poisson}(Z_i\lambda) \text{ and } Z_i \sim \text{Bernoulli}(p)$$

where Z_i is the latent indicator that visitor i fished and thus the mean of Y_i is zero for non-fishers with $Z_i = 0$. In this scenario, we do not observe the Z_i but this two-stage model gives the equivalent model to (4.31) but uses only standard distributions. As a result, the model can be coded in JAGS with covariates included in the mass at zero and the Poisson rate as in Listing 4.7e.

4.3.3 Example: Logistic regression for NBA clutch free throws

The table in Problem 17 of Section 1.6 gives the overall free-throw proportion ($q_i \in [0, 1]$), the number of made clutch shots (Y_i), and the number of attempted clutch shots (n_i) for players $i = 1, ..., 10$. Figure 4.4 shows that most players have similar success rates in clutch and non-clutch situations, but some players appear to have less success in pressure situations. We fit logistic regression models to formally explore this relationship.

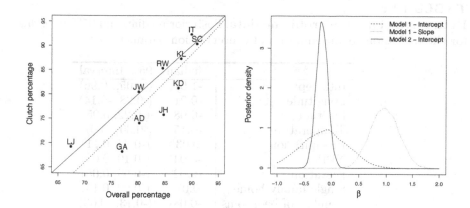

FIGURE 4.4

Logistic regression analysis of NBA free throws. The first panel shows the overall percentage of made free throws versus the percentage for clutch shots only for each player (denoted by the player's initials). The solid line is the $x = y$ lines and the dashed line is the fitted value from Model 2. The second plot is the posterior density for the slope and intercept from the Model 1 and the intercept from the model 2.

The support of the response is $Y_i \in \{0, 1, ..., n_i\}$ and so we select a binomial likelihood $Y_i|p_i \sim \text{Binomial}(n_i, p_i)$ where p_i is the probability of player i making a clutch shot. The number of shots n_i is known, and so we link the covariate to the success probability p_i. The two models are

1. $\text{logit}(p_i) = \beta_1 + \beta_2 X_i$
2. $\text{logit}(p_i) = \beta_1 + X_i$

where $X_i = \text{logit}(q_i)$ is the log odds of making a regular free throw. If $\beta_1 = 0$ and $\beta_2 = 1$ in Model 1 or $\beta_1 = 0$ in Model 2, then the clutch performance equals the overall performance. To determine if this is the case, we fit the model using JAGS with two chains with burn-in 10,000 and 20,000 additional samples and with uninformative Normal$(0, 100)$ priors for all parameters. The model specification for Model 1 is

```
for(i in 1:10){
  Y[i]       ~ dbinom(p[i],n[i])
  logit(p[i]) <- beta[1] + beta[2]*X[i]
}
beta[1] ~ dnorm(0,0.01)
beta[2] ~ dnorm(0,0.01)
```

The results are plotted in Figure 4.4. For Model 1, the slope is centered squarely on one and the intercept is centered slightly below zero, but both

TABLE 4.1
Beta regression of microbiome data. Posterior median and 95% intervals
for the regression coefficients β_j and concentration parameter r.

	Median	95% Interval
Intercept	-1.01	(-1.05, -0.96)
Longitude	-0.21	(-0.28, -0.14)
Latitude	-0.08	(-0.22, 0.05)
Temperature	-0.15	(-0.30, -0.01)
Precipitation	0.03	(-0.04, 0.11)
NPP	-0.04	(-0.10, 0.02)
Elevation	-0.02	(-0.09, 0.05)
Single-family home	0.07	(0.02, 0.13)
Number of bedrooms	-0.08	(-0.13, -0.03)
r	7.97	(7.34, 8.66)

parameters have considerable uncertainty and so the model with $p_i = q_i$ remains plausible. However, the intercept in Model 2 is negative with posterior probability 0.96, so there is some evidence that even the best players in the NBA underperform in pressure situations. The fitted curve (dashed line) in Figure 4.4 (left panel) is $1/[1+\exp(-\bar{\beta}_1 - X_i)]$, where $\bar{\beta}_1$ is the posterior mean, and shows that the clutch performance can be several percentage points lower than the overall percentage.

4.3.4   Example: Beta regression for microbiome data

In Section 4.2.5 we regressed the diversity of a sample's microbiome onto features of the home. Diversity was measured as the log of the number of the $L = 763$ species present in the sample. However, this measure fails to account for the relative abundance of the species. Let $A_{il} \geq 0$ be the abundance of species l in sample i. Another measure of the diversity is the proportion of the total abundance attributed to the most abundant species in sample i,

$$Y_i = \frac{\max\{A_{i1}, ..., A_{iL}\}}{\sum_{l=1}^{L} A_{il}} \in [0, 1]. \tag{4.32}$$

This measure is plotted in Figure 4.5 (top left).

Since Y_i is between 0 and 1, Gaussian linear regression is inappropriate. One option is to transform the response from support $[0, 1]$ to $(-\infty, \infty)$, e.g., $Y_i^* = \text{logit}(Y_i)$, and model the transformed data Y_i^* using a Gaussian linear model. Another option is to model the data directly using a non-Gaussian model. A natural model for a continuous variable with support $[0, 1]$ is the beta distribution. Since the mean must also be in $[0, 1]$, a logistic link can be used to relate the covariates to the mean response. Therefore, we fit the beta regression model

$$Y_i|\boldsymbol{\beta}, r \sim \text{Beta}[rq_i, r(1 - q_i)] \text{ and logit}(q_i) = \mathbf{X}_i^T\boldsymbol{\beta}. \quad (4.33)$$

$Y_i|\boldsymbol{\beta}, r$ has mean q_i and variance $q_i(1 - q_i)/(r + 1)$, and so $r > 0$ determines the concentration of the beta distribution around the mean q_i.

Using priors $\beta_j \sim \text{Normal}(0, 100)$ and $r \sim \text{Gamma}(0.1, 0.1)$, we fit this model in JAGS using the code in Listing 4.7e (although sampling would likely be faster using the **betareg** package in R). Before fitting the model, the covariates are standardized to have mean zero and variance one. Convergence was excellent for all parameters (the bottom left panel of Figure 4.5 shows the trace plot for r). The posterior distributions in Table 4.1 indicate that there is more diversity (smaller Y_i) on average in homes in cool regions in the east, and multiple-family homes with many bedrooms.

4.4 Random effects

The standard linear regression model assumes the same regression model applies to all observations. This assumption is tenuous if data are collected in groups. For example, Figure 4.6 plots the jaw bone density measurements of $n = 20$ children measured over the course of $m = 4$ visits. These 20 children represent only a random sample from a much larger population, but we can use these samples to make an inference about the larger population. If we let Y_{ij} be the jth measurement for patient i, then (ignoring age as in the top right panel of Figure 4.6) the one-way random effects model is

$$Y_{ij}|\alpha_i \stackrel{indep}{\sim} \text{Normal}(\alpha_i, \sigma^2) \text{ where } \alpha_i \stackrel{iid}{\sim} \text{Normal}(\mu, \tau^2). \quad (4.34)$$

The random effect α_i is the true mean for patient i, and the observations for patient i vary around α_i with variance σ^2. The α_i are called random effects because if we repeated the experiment with a new sample of 20 children the α_i would change. In this model, the population of patient-specific means is assumed to follow a normal distribution with mean μ and variance τ^2. The overall mean μ is a fixed effect because if we repeated the experiment with a new sample of 20 children from the same population it would not change. A linear model with both fixed and random effects is called a linear mixed model.

A Bayesian analysis of a random effects model requires priors for the population parameters. For example, Listing 4.8 provides JAGS code with conjugate priors $\mu \sim \text{Normal}(0, 100^2)$ and $\tau^2 \sim \text{InvGamma}(0.1, 0.1)$. The same algorithms (e.g., Gibbs sampling) can be used for random effects models as for the other models we have considered. In fact, computationally there is no need to distinguish between fixed and random effects (which can lead to

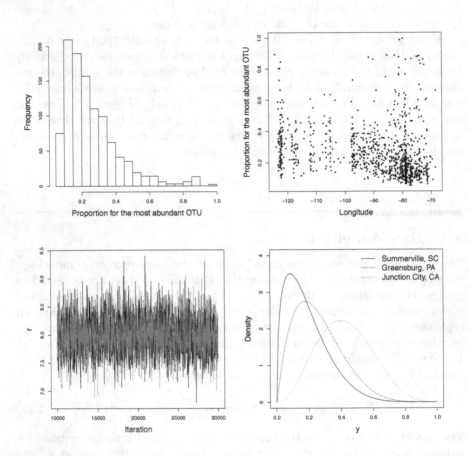

FIGURE 4.5

Beta regression for microbiome data. The top left panel shows the histogram of the observed proportions of abundance allocated to the most abundance OTU, and the top right panel plots this variable against the sample's longitude. The second row gives the trace plots (the two chains are different shades of gray) of the concentration parameter, r, and the fitted Beta$[\hat{r}\hat{q}, \hat{r}(1 - \hat{q})]$ density for three samples evaluated at the posterior mean for all parameters.

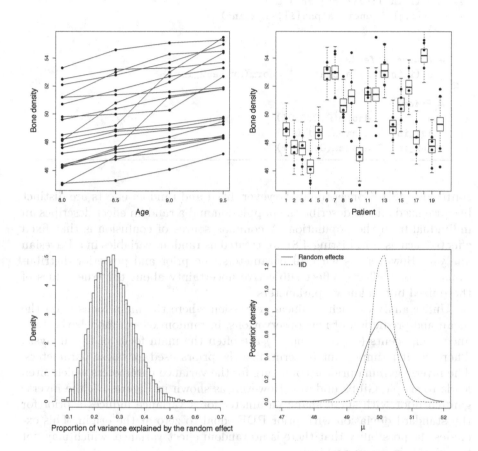

FIGURE 4.6
One-way random effect analysis of the jaw data. The dots in the top
left panel show bone density at the four visits for each patient (connected by
lines), the top right panel compares the observations (dots) and the posterior
distribution of the subject random effect α_i (boxplot), the bottom left panel
plots the posterior of the variance ratio $\tau^2/(\tau^2 + \sigma^2)$, and the final panel
compares the posterior density of the mean μ from the random effects model
versus independence model $Y_{ij} \overset{iid}{\sim} \text{Normal}(\mu, \sigma^2)$.

Listing 4.8
The one-way random effects model in JAGS.

```
1   # Likelihood
2   for(i in 1:n){for(j in 1:m){
3     Y[i,j] ~ dnorm(alpha[i],sig2_inv)
4   }}
5
6   # Random effects
7   for(i in 1:n){alpha[i] ~ dnorm(mu,tau2_inv)}
8
9   # Priors
10  mu        ~ dnorm(0,0.0001)
11  sig2_inv ~ dgamma(0.1,0.1)
12  tau2_inv ~ dgamma(0.1,0.1)
```

confusion, [42]). Conceptually, however, fixed and random effects are distinct because fixed effects describe the population and a random effect describes an individual from the population. A common source of confusion is that fixed effects (such as μ in Listing 4.8) are treated as random variables in a Bayesian analysis. However, as with all parameters, the prior and posterior distributions for fixed effects reflect subjective uncertainty about the true values of these fixed but unknown parameters.

Unlike analyses such as linear regression where the main focus is on the mean and prediction of new observations, in random effects models the variance components (e.g., σ^2 and τ^2) are often the main focus of the analysis. Therefore it is important to scrutinize the priors used for these parameters. The inverse gamma prior is conjugate for the variance parameters which often leads to simple Gibbs updates. However, as shown in Figure 4.7, the inverse gamma prior with small shape parameter for a variance induces a prior for the standard deviation with prior PDF equal to zero at the origin. This excludes the possibility that there is no random effect variance, which may not be suitable in many problems.

As an alternative, [29] endorses a half-Cauchy (HC) prior for the standard deviation. The HC distribution is the student-t distribution with one degree of freedom restricted to be positive and has a flat PDF at the origin (Figure 4.7), which is usually a more accurate expression of prior belief. Listing 4.9 gives JAGS code for this prior. In this code the HC distribution is assigned directly to the standard deviations which breaks the conjugacy relationships for the variance components (this is easily handled by JAGS); [29] shows that conjugacy can be restored using a two-stage model. This code assumes the HC scale parameter is fixed at one. Because the Cauchy prior has a very heavy tail this gives 0.99 prior quantile equal to 63. Despite this wide prior range, the scale of the HC prior should be adjusted to the scale of the data.

Random effects induce correlation between observations from the same group. In the one-way random effects model, the covariance marginally over

Listing 4.9
The one-way random effects model with half-Cauchy priors.

```
 1   # Likelihood
 2   for(i in 1:n){for(j in 1:m){
 3     Y[i,j] ~ dnorm(alpha[i],sig2_inv)
 4   }}
 5
 6   # Random effects
 7   for(i in 1:n){alpha[i] ~ dnorm(mu,tau2_inv)}
 8
 9   # Priors
10   mu          ~ dnorm(0,0.0001)
11   sig2_inv <- pow(sigma1,-2)
12   tau2_inv <- pow(sigma2,-2)
13   sigma1      ~ dt(0, 1, 1)T(0,) # Half-Cauchy priors with
14   sigma2      ~ dt(0, 1, 1)T(0,) # location 0 and scale 1
```

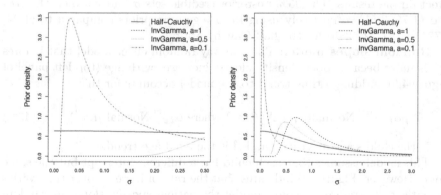

FIGURE 4.7
Priors for a standard deviation. The half-Cauchy prior for σ and the prior induced for σ by inverse gamma priors on σ^2 with different shape parameters. All priors are scaled to have median equal 1. The two panels differ only by the range of σ being plotted.

the random effect α_i is

$$\text{Cov}(Y_{ij}, Y_{uv}) = \begin{cases} \tau^2 + \sigma^2 & i = u \text{ and } j = v \\ \tau^2 & i = u \text{ and } j \neq v \\ 0 & i \neq u. \end{cases} \quad (4.35)$$

The variance of each observation is $\tau^2 + \sigma^2$ and includes both the variance in the random effect (τ^2) and variability around the mean (σ^2). Since two observations from the same patient share a common random effect, they have covariance τ^2 and thus correlation $\tau^2/(\tau^2 + \sigma^2)$. Observations from different patients have no common source of variability and are thus uncorrelated.

Returning to the bone density data, Figure 4.6 plots the posterior approximated with two chains each with 30,000 samples and the first 10,000 discarded as burn-in assuming the model in Listing 4.8. The posterior distribution of the proportion of the variance attributed to the random effect $\tau^2/(\tau^2 + \sigma^2)$ ranges from 0.1 to 0.5 (bottom left panel). Accounting for this correlation between observations from the same patient affects the posterior of the fixed effect, μ (bottom right). The posterior variance of μ is larger for the random effects model compared to the posterior of μ for the independence model $Y_{ij} \overset{iid}{\sim} \text{Normal}(\mu, \sigma^2)$. This is expected because there is less information about the mean in 4 repeated measurements for 20 patients than 80 measurements from 80 different patients.

To test for prior sensitivity, we refit the model using a half-Cauchy prior in Listing 4.9. In this case the prior for the variance components had little effect on the results. The 95% posterior credible sets σ and τ are (1.24, 1.78) and (1.75, 3.51) respectively using inverse gamma priors compared to (1.24, 1.77) and (1.72, 3.45) for the half-Cauchy priors.

Random slopes model: The one-way random effect model that ignores age is naive because bone density clearly increases with age (top left panel of Figure 4.6). Adding a time trend to the model accounts for this,

$$Y_{ij}|\alpha_i \overset{indep}{\sim} \text{Normal}(\alpha_i + X_j\beta, \sigma^2) \text{ where } \alpha_i \overset{iid}{\sim} \text{Normal}(\mu, \tau^2), \quad (4.36)$$

X_j is the child's age at visit j, and β is the fixed age trend.

This model assumes that each child has a different intercept but the same slope. However, Figure 4.6 indicates that the rate of increase over time varies by patient. Therefore we could model the patient-specific slopes as random effects

$$Y_{ij}|\boldsymbol{\alpha}_i \overset{indep}{\sim} \text{Normal}(\alpha_{i1} + \alpha_{i2}X_j, \sigma^2) \text{ where } \boldsymbol{\alpha}_i \overset{iid}{\sim} \text{Normal}(\boldsymbol{\beta}, \boldsymbol{\Omega}), \quad (4.37)$$

and the random intercept and slope for patient i is $\boldsymbol{\alpha}_i = (\alpha_{i1}, \alpha_{i2})^T$. The mean vector $\boldsymbol{\beta}$ includes the population mean intercept and slope, and is thus a fixed effect. The 2×2 population covariance matrix $\boldsymbol{\Omega}$ determines the variation of the random effects over the population. To complete the Bayesian model we specify prior $\sigma^2 \sim \text{InvGamma}(0.1, 0.1)$, $\boldsymbol{\beta} \sim \text{Normal}(\mathbf{0}, 100^2\mathbf{I}_2)$, and $\boldsymbol{\Omega} \sim$

Listing 4.10
Random slopes model in JAGS.

```
1    # Likelihood
2      for(i in 1:n){for(j in 1:m){
3        Y[i,j] ~ dnorm(alpha[i,1]+alpha[i,2]*age[j],tau)
4      }}
5
6    # Random effects
7      for(i in 1:n){
8        alpha[i,1:2] ~ dmnorm(beta[1:2],Omega_inv[1:2,1:2])
9      }
10
11   # Priors
12     tau ~ dgamma(0.1,0.1)
13     for(j in 1:2){beta[j] ~ dnorm(0,0.0001)}
14     Omega_inv[1:2,1:2] ~ dwish(R[,],2.1)
15
16     R[1,1]<-1/2.1
17     R[1,2]<-0
18     R[2,1]<-0
19     R[2,2]<-1/2.1
```

InvWishart$(2.1, \mathbf{I}_2/2.1)$. The inverse Wishart prior for the covariance matrix has prior mean $\mathbf{I}_2$, the 2×2 identify matrix (see Appendix A.1). JAGS code for this random-slopes model is in Listing 4.10.

The posterior mean of the population covariance matrix is

$$E(\mathbf{\Omega}|\mathbf{Y}) = \begin{bmatrix} 91.78 & -10.14 \\ -10.14 & 1.23 \end{bmatrix} \tag{4.38}$$

and the posterior 95% interval for the correlation $\mathrm{Cor}(\alpha_{i1}, \alpha_{i2}) = \Omega_{12}/\sqrt{\Omega_{11}\Omega_{22}}$ is (-0.98, -0.89). Therefore there is a strong negative dependence between the intercept and slope, indicating that bone density increases rapidly for children with low bone density at age 8, and vice versa.

Figure 4.8 plots the posterior distribution of the fitted values $\alpha_{i1} + \alpha_{i2}X$ for X between 8 and 10 years for three patients. For each patient and each age, we compute the 95% interval using the quantiles of the S posterior samples $\alpha_{i1}^{(s)} + \alpha_{i2}^{(s)}X$. We also plot the posterior predictive distribution (PPD) for the measured bone density at age 10. The PPD is approximated by sampling $Y_i^{*(s)} \sim \mathrm{Normal}(\alpha_{i1}^{(s)} + \alpha_{i2}^{(s)}10, \sigma^{2(s)})$ at each iteration and then computing the quantiles of the S predictions. The PPD accounts for both uncertainty in the patient's random effect $\boldsymbol{\alpha}_i$ and measurement error with variance σ^2. The intervals in Figure 4.8 suggest that uncertainty in the random effects is the dominant source of variation.

Marginal models: Inducing correlation by conditioning on random effects is equivalent to a marginal model (Section 4.5.4) that does not include random

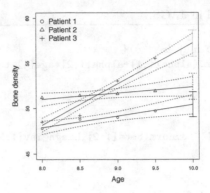

FIGURE 4.8
Mixed effects analysis of the jaw data. The observed bone density (points) for three subjects versus the posterior median (solid lines) and 95% intervals (dashed lines) of the fitted value $\alpha_{i1} + \alpha_{i2}X$ for X ranging from 8–10 years, and 95% credible intervals (vertical lines at Age=10) of the posterior predictive distributions for the measured response at age $X = 10$.

effects but directly specifies correlation between observation in the same group. For example, the one-way random effects model

$$Y_{ij}|\alpha_i \overset{indep}{\sim} \text{Normal}(\alpha_i, \sigma^2) \text{ where } \alpha_i \overset{iid}{\sim} \text{Normal}(\mu, \tau^2) \qquad (4.39)$$

is equivalent to the marginal model

$$\mathbf{Y}_i \sim \text{Normal}(\boldsymbol{\mu}, \boldsymbol{\Sigma}), \qquad (4.40)$$

where $\mathbf{Y}_i = (Y_{i1}, ..., Y_{im})^T$ is the data vector for group i, $\boldsymbol{\mu} = (\mu, ..., \mu)^T$ is the mean vector, and $\boldsymbol{\Sigma}$ is the covariance matrix with $\tau^2 + \sigma^2$ on the diagonals and τ^2 elsewhere. An advantage of the marginal approach is that we no longer have to estimate the random effects α_i; a disadvantage is that for large data sets and complex correlation structure the mean vector and especially the covariance matrix can be large which slows computation.

In the hierarchical representation, the elements of $\mathbf{Y}_i$ are independent and identically distributed given α_i; in the marginal model the m observations from group i are no longer independent, but they remain exchangeable, i.e., their distribution is invariant to permuting their order. The concept of exchangeability plays a fundamental role in constructing hierarchical models. The representation theorem by Bruno de Finetti states that any infinite sequence of exchangeable variables can be written as independent and identically distributed conditioned on some latent distribution. Therefore, this important type of dependent data can be modeled using a simpler hierarchical model.

4.5 Flexible linear models

As discussed in Section 4.2, the multiple linear regression model of response Y_i onto covariates $X_{i1}, ..., X_{ip}$ is

$$Y_i = \sum_{j=1}^{p} X_{ij}\beta_j + \varepsilon_i, \qquad (4.41)$$

where $\boldsymbol{\beta} = (\beta_1, ..., \beta_p)^T$ are the regression coefficients and the errors are $\varepsilon_i \overset{iid}{\sim}$ Normal$(0, \sigma^2)$. This model makes four key assumptions:

(1) Linearity: The mean of $Y_i|\mathbf{X}_i$ is linear in $\mathbf{X}_i$

(2) Equal variance: The residual variance (σ^2) is the same for all i

(3) Normality: The errors ε_i are Gaussian

(4) Independence: The errors ε_i are independent

In real analyses, most if not all of these assumptions will be violated to some extent. Minor violations will not invalidate statistical inference, but glaring model misspecifications should be addressed. This chapter provides Bayesian remedies to model misspecification (each subsection addresses one of the four assumptions above). A strength of the Bayesian paradigm is that these models can be fit by simply adding a few lines of JAGS code and do not require fundamentally new theory or algorithms.

4.5.1 Nonparametric regression

The linearity assumption can be relaxed by using a general expression of the regression of Y_i onto $\mathbf{X}_i$,

$$Y_i = g(\mathbf{X}_i) + \varepsilon_i = g(X_{i1}, ..., X_{ip}) + \varepsilon_i \qquad (4.42)$$

where g is the mean function and $\varepsilon_i \overset{iid}{\sim}$ Normal$(0, \sigma^2)$. Parametric regression specifies the mean function g as a parametric function of a finite number of parameters, e.g., in linear regression $g(\mathbf{X}_i) = \mathbf{X}_i\boldsymbol{\beta}$. A linear mean function is often a sufficient and interpretable first-order approximation, but more complex relationships between variables can be fit using a more flexible model.

For example, consider the data from the mcycle R package plotted in Figure 4.9. The predictor, X_i, is the time since a motorcycle makes impact (scaled to the unit interval) and the response, Y_i, is the acceleration of a monitor on the head of a crash test dummy. Clearly a linear model will not fit these data well: the mean is flat for the first quarter of the experiment, then dramatically dips, rebounds, and then plateaus until the end of the experiment.

A fully nonparametric model allows for any continuous function g. A model

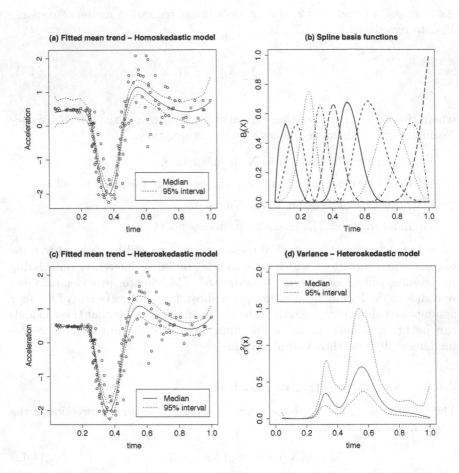

FIGURE 4.9
Nonparametric regression for the motorcycle data. Panel (a) plots the time since impact (scaled to be between 0 and 1) and the acceleration (g) along with the posterior median and 95% interval for the mean function from the homoskedastic fit; Panel (b) shows the $J = 10$ spline basis functions $B_j(X)$; Panels (c) and (d) show the posterior median and 95% intervals for the mean and variance functions from the heteroskedastic model.

this flexible requires infinitely many parameters. We will focus on semiparametric models that specify the mean function in terms of a finite number of parameters in a way that increasing the number of parameters can approximate any function g. For example, if there is only $p = 1$ covariate (X) we could fit a Jth order polynomial function

$$g(X) = \sum_{j=0}^{J} X^j \beta_j. \tag{4.43}$$

This model has $J+1$ parameters and by increasing J the polynomial function can approximate any continuous g.

There are many Bayesian semiparametric/nonparametric regression models, including Gaussian process regression [67], Bayesian adaptive regression trees [18], neural networks [61] and regression splines [19]. The simplest approach is arguably regression splines. In spline regression we construct non-linear functions of the original covariates, and use these constructed covariates as the predictors in multiple linear regression. Denote the J constructed covariates as $B_1(\mathbf{X}), ..., B_J(\mathbf{X})$. In polynomial regression $B_j(X) = X^j$, but there are many other choices. For example, Figure 4.9b plots $J = 10$ B-spline basis functions. These functions are appealing because they are smooth and local, i.e., non-zero only for some values of X. The model is simply the multiple linear regression model (Section 4.2) with the $B_1(X), ..., B_J(X)$ as the covariates,

$$Y_i \sim \text{Normal}[g(X_i), \sigma^2] \text{ and } g(X_i) = \beta_0 + \sum_{j=1}^{J} B_j(X_i)\beta_j.$$

Note that each basis function in Figure 4.9b has $B_j(0) = 0$, and so an intercept (β_0) is required. By increasing J, any smooth mean function can be approximated as a linear combination of the B-spline basis functions.

Motorcycle example: To fit the mean curve to the data plotted in Figure 4.9a, we use $J = 10$ B-spline basis functions and priors $\beta_j \sim \text{Normal}(0, \tau^2\sigma^2)$ and $\sigma^2, \tau^2 \sim \text{InvGamma}(0.1, 0.1)$. The model is fit using MCMC with the code in Listing 4.3. In this code, the basis functions have been computed in using the bs package in R and passed to JAGS as $x_{ij} = B_j(X_i)$. For each iteration, we compute $g(X_i)$ for all $i = 1, ..., n$ as a function of that iteration's posterior sample of $\boldsymbol{\beta}$. This produces the entire posterior distribution of the mean function g for all n sample points (and any other X we desire), and Figure 4.9a plots the posterior median and 95% interval of g at each sample point. The fitted model accurately captures the main trend including the valley around $X = 0.4$.

We selected $J = 10$ basis functions because this degree of model complexity visually seemed to fit the data well. Choosing smaller J would give a smoother estimate of g and choosing larger J would give a rougher estimate. Clearly a more rigorous approach to selecting the number of basis functions is needed, and this is discussed in Chapter 5.

Listing 4.11
Model statement for heteroskedastic Gaussian regression.

```
1    for(i in 1:n){
2      Y[i]    ~ dnorm(mu[i],prec[i])
3      mu[i]   <- inprod(x[i,],beta[])
4      prec[i] <- 1/sig2[i]
5      sig2[i] <- exp(inprod(x[i,],alpha[]))
6    }
7    for(j in 1:p){beta[j] ~ dnorm(0,taub)}
8    for(j in 1:p){alpha[j] ~ dnorm(0,taua)}
9    taub ~ dgamma(0.1,0.1)
10   taua ~ dgamma(0.1,0.1)
```

4.5.2   Heteroskedastic models

The standard linear regression analysis assumes a homoskedastic variance $V(\varepsilon_i) = \sigma^2$ for all i. A more flexible heteroskedastic model allows the covariates to affect both the mean and the variance. A natural approach is to build a linear model for the variance as a function of the covariates so that $V(\varepsilon_i) = \sigma^2(\mathbf{X}_i)$. Since the variance is positive, we must transform the linear predictor to be positive before linking to the variance. For example,

$$\log[\sigma^2(\mathbf{X}_i)] = \sum_{j=1}^{p} X_{ij}\alpha_j, \qquad (4.44)$$

or equivalently $\sigma^2(\mathbf{X}_i) = \exp(\sum_{j=1}^{p} X_{ij}\alpha_j)$. The parameter α_j determines the effect of the covariate j on the variance and must be estimated. JAGS code for this model is in Listing 4.11.

Motorcycle example: The variance of the observations about the mean trend in Figure 4.9a clearly depends on X, with small variance at the beginning of the experiment and large variance in the middle. To capture this heteroskedasticity in a flexible way, we model the log variance using the same J B-spline basis functions used for the mean,

$$g(X) = \beta_0 + \sum_{j=1}^{p} B_j(X)\beta_j \text{ and } \log[\sigma^2(X)] = \alpha_0 + \sum_{j=1}^{p} B_j(X)\alpha_j, \qquad (4.45)$$

where $\beta_j \sim \text{Normal}(0, \sigma_b^2)$ and $\alpha_j \sim \text{Normal}(0, \sigma_a^2)$. The hyperparameters have uninformative priors $\sigma_a^2, \sigma_b^2 \sim \text{InvGamma}(0.1, 0.1)$.

The pointwise 95% intervals of $\sigma^2(X)$ in Figure 4.9d suggest that the variance is indeed small at the beginning of the experiment and increases with X. Comparing the posterior distributions of the mean trend for the homoskedastic (Figure 4.9a) and heteroskedastic (Figure 4.9c) models, the posterior means are similar but the heteroskedastic model produces more realistic 95% intervals with widths that vary according the pattern of the error variance. Therefore, it appears that properly quantifying uncertainty about the mean trend requires a realistic model for the error variance.

4.5.3 Non-Gaussian error models

Most Bayesian regression models assume Gaussian errors, but more flexible methods are easily constructed. For example, to accommodate heavy tails the errors could be modelled using a student-t or double-exponential (Laplace) distributions. To further allow for asymmetry, generalizations such as the skew-t and asymmetric Laplace distributions would be used. Listing 4.12a provides code for regression with student-t errors.

In most analysis, a suitable parametric distribution can be found. However, this process is subjective and difficult to automate. Just as a mixture of conjugate priors (Section 2.1.8) can be used to approximate virtually any prior distribution, the mixture of normals distribution can be used to approximate virtually any residual distribution. The mixture of normals density for ε is

$$f(\varepsilon) = \sum_{k=1}^{K} \pi_k \phi(\varepsilon; \theta_k, \tau_1^2), \qquad (4.46)$$

where K is the number of mixture components, $\pi_k \in (0, 1)$ is the probability on mixture component k, and $\phi(\varepsilon; \theta, \tau^2)$ is the Gaussian PDF with mean θ and variance τ^2. This model is equivalent to the clustering model where

$$Y_i | g_i \sim \text{Normal}\left(\sum_{j=1}^{p} X_{ij}\beta_j + \theta_{g_i}, \sigma^2\right) \qquad (4.47)$$

and $g_i \in \{1, ..., K\}$ is the cluster label for observation i with $\text{Prob}(g_i = k) = \pi_k$. By letting the number of mixture components increase to infinity, any distribution can be approximated, and by selecting priors $\theta_k \overset{iid}{\sim} \text{Normal}(0, \tau_2^2)$, $\tau_1^2, \tau_2^2 \sim \text{InvGamma}$, and $(\pi_1, ..., \pi_K) \sim \text{Dirichlet}$ all full conditional distributions for all parameters are conjugate permitting Gibbs samples (Listing 4.12b).

For a fixed number of mixture components (K) the mixture-of-normals model is a semiparametric estimator of the density $f(\varepsilon)$. There is a rich literature on nonparametric Bayesian density estimation [37]. The most common model is the Dirichlet process mixture model that has infinitely many mixture components and a particular model for the mixture probabilities.

4.5.4 Linear models with correlated data

For data with a natural ordering such as spatial or temporal data modeling the correlation between observations is important to obtain valid inference and make accurate predictions. Generally, the Bayesian linear model with

Listing 4.12
Model statement for Gaussian regression with non-normal errors.

```
1
2    # (a) Regression with student-t errors
3    for(i in 1:n){
4      Y[i]   ~ dt(mu[i],tau,df)
5      mu[i] <- inprod(X[i,],beta[])
6    }
7    for(j in 1:p){beta[j] ~ dnorm(0,taub)}
8    tau ~ dgamma(0.1,0.1)
9    df  ~ dgamma(0.1,0.1)
10
11
12   # (b) Regression with mixture-of-normals errors
13   for(i in 1:n){
14     Y[i]   ~ dnorm(mu[i]+theta[g[i]],tau1)
15     mu[i] <- inprod(X[i,],beta[])
16     g[i]   ~ dcat(pi[])
17   }
18   for(k in 1:K){theta[k] ~ dnorm(0,tau2)}
19   for(j in 1:p){beta[j] ~ dnorm(0,tau3)}
20   tau1   ~ dgamma(0.1,0.1)
21   tau2   ~ dgamma(0.1,0.1)
22   tau3   ~ dgamma(0.1,0.1)
23   pi[1:K] ~ ddirch(alpha[1:K])
```

correlated errors is

$$\mathbf{Y} \sim \text{Normal}(\mathbf{X}\boldsymbol{\beta}, \boldsymbol{\Sigma}). \tag{4.48}$$

A Bayesian analysis of correlated data hinges on correctly specifying the correlation structure to capture say spatial or temporal correlation. Given the correlation structure and priors for the correlation parameters, standard Bayesian computational tools can be used to summarize the posterior. Correlation parameters usually will not have conjugate priors and so Metropolis–Hastings sampling is used. An advantage of the Bayesian approach for correlated data is that using MCMC sampling we can account for uncertainty in the correlation parameters for prediction or inference on other parameters, whereas maximum likelihood analysis often uses plug-in estimates of the correlation parameters and thus underestimates uncertainty.

Gun control example: The data for this analysis come from Kalesan et. al. (2016) [47]. The response variable, Y_i, is the log firearm-related death rate per 10,000 people in 2010 in state i (excluding Alaska and Hawaii). This is regressed onto five potential confounders: log 2009 firearm death rate per 10,000 people; firearm ownership rate quartile; unemployment rate quartile; non-firearm homicide rate quartile; and firearm export rate quartile. The covariate of interest is the number of gun control laws in effect in the state. This gives $p = 6$ covariates.

We first fit the usual Bayesian linear regression model

$$Y_i = \beta_0 + \sum_{j=1}^{p} X_i \beta_j + \varepsilon_i \tag{4.49}$$

with independent errors $\varepsilon_i \overset{iid}{\sim} \text{Normal}(0, \sigma^2)$ and uninformative priors. The posterior density of the regression coefficient corresponding to the number of gun laws is plotted in Figure 4.10. The posterior probability that the coefficient is negative is 0.96, suggesting a negative relationship between the number of gun laws and the firearm-related death rate.

The assumption of independent residuals is questionable because neighboring states may be correlated. Spatial correlation may stem from guns being brought across state borders or from missing covariates (e.g., attitudes about and use of guns) that vary spatially. Research has shown that accounting for residual dependence can have a dramatic effect on regression coefficient estimates [43].

We decompose the residual covariance $\text{Cov}[(\varepsilon_1, ..., \varepsilon_n)^T] = \Sigma$ as

$$\Sigma = \tau^2 S + \sigma^2 I_n, \tag{4.50}$$

where $\tau^2 S$ is the spatial covariance and $\sigma^2 I_n$ is the non-spatial covariance. There are many spatial correlation models (e.g., [4]) that allow the correlation between two states to decay with the distance between the states. For example, a common model is to assume the correlation between states decays

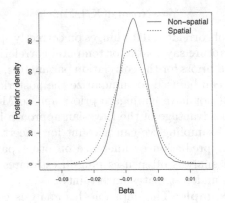

FIGURE 4.10
Effect of gun-control legislation on firearm-related death rate. Posterior distribution of the coefficient associated with the number of gun-control laws in a state from the spatial and non-spatial model of the states' firearm-related death rate.

exponentially with the distance between them. However, quantifying the distance between irregularly shaped states is challenging, and so we model spatial dependence using adjacencies. Let $A_{ij} = 1$ if states i and j share a border and $A_{ij} = 0$ if $i = j$ or the states are not neighbors. The spatial covariance follows the conditionally autoregressive model $S = (M - \rho A)^{-1}$, where A is the adjacency matrix with (i, j) element A_{ij} and M is the diagonal matrix with the ith diagonal element equal to the number of states that neighbor state i. The parameter $\rho \in (0, 1)$ is not the correlation between adjacent sites, but determines the strength of spatial dependence with $\rho = 0$ corresponding to independence.

The posterior mean (standard deviation) of the spatial dependence parameter ρ is 0.38 (0.25), and so the residual spatial dependence in these data is not strong. However, the posterior of the regression coefficient of interest in Figure 4.10 is noticeably wider for the spatial model than the non-spatial model. The posterior probability that the coefficient is negative lowers from 0.96 from the non-spatial model to 0.93 for the spatial model. Therefore, while accounting for residual dependence did not qualitatively change the results, this example illustrates that the chosen model for the residuals can affect the posterior of the regression coefficients.

Jaw bone density example: A possible correlation structure for the longitudinal data in Figure 4.6 (top left) is to assume that correlation decays with the time between visits. A first-order autoregression correlation structure is $\text{Cor}(Y_{ij}, Y_{ik}) = \rho^{|j-k|}$. Denoting the vector of m observations for patient

Listing 4.13
Random slopes model with autoregressive dependence in JAGS.

```
1   # Likelihood
2     for(i in 1:n){
3       Y[i,1:m] ~ dmnorm(mn[i,1:m],SigmaInv)
4       for(j in 1:m){mn[i,j] <- alpha[i,1]+alpha[i,2]*age[j]}
5     }
6     SigmaInv[1:m,1:m] <- inverse(Sigma[1:m,1:m])
7     for(j in 1:m){for(k in 1:m){
8       Sigma[j,k] <- pow(rho,abs(k-j))/tau
9     }}
10
11  # Random effects
12    for(i in 1:n){alpha[i,1:2] ~ dmnorm(beta[1:2],Omega[1:2,1:2])}
13
14  # Priors
15    tau ~ dgamma(0.1,0.1)
16    for(j in 1:2){beta[j] ~ dnorm(0,0.0001)}
17    rho ~ dunif(0,1)
18    Omega[1:2,1:2] ~ dwish(R[,],2.1)
19
20    R[1,1]<-1/2.1
21    R[1,2]<-0
22    R[2,1]<-0
23    R[2,2]<-1/2.1
```

i as $\mathbf{Y}_i = (Y_{i1}, ..., Y_{im})^T$, the $m \times m$ covariance matrix $\boldsymbol{\Sigma}$ has (j, k) element equal to $\sigma^2 \rho^{|j-k|}$. The random slope model for the mean in matrix notation is $E(\mathbf{Y}_i | \boldsymbol{\alpha}_i) = \mathbf{X}\boldsymbol{\alpha}_i$, where $\mathbf{X}$ is the $m \times 2$ matrix with the first column equal to the vector of ones for the intercept and the second column equal to the ages $X_1, ..., X_m$ for the slope. The likelihood is then

$$\mathbf{Y}_i | \alpha_i \overset{indep}{\sim} \text{Normal}(\mathbf{X}\alpha_i, \boldsymbol{\Sigma}) \qquad (4.51)$$

with random effects distribution $\alpha_i \overset{iid}{\sim} \text{Normal}(\boldsymbol{\beta}, \boldsymbol{\Omega})$. The correlation parameter is given prior $\rho \sim \text{Uniform}(0, 1)$ and all other priors are the same as other fits. JAGS code is given in Listing 4.13.

The posterior median of the correlation parameter ρ is 0.85 and the posterior 95% interval is (0.46, 0.96) so there is evidence of correlation that cannot be explained by the patient-specific linear trend. Including autoregressive correlation has only a modest effect on the posterior distribution of the fixed effects: the 95% posterior intervals for the random effect model with independent errors are (29.9, 38.3) for β_1 and (1.33, 2.38) for β_2 compared to (30.0, 37.3) for β_1 and (1.45, 2.31) for β_2 for the autoregressive model.

This model with random intercept, random slope and autoregressive correlation structure is now very complex. There are almost as many parameters as observations and multiple explanations of dependence (random effects and residual correlation). Perhaps in this case all of these terms are necessary and can be estimated from this relatively small data set, but a simpler yet adequate model is preferred for computational purposes and because simpler models are easier to explain and defend. Model comparisons and tests of model adequacy are the topics of Chapter 5.

4.6 Exercises

1. A clinical trial gave six subjects a placebo and six subjects a new weight loss medication. The response variable is the change in weight (pounds) from baseline (so -2.0 means the subject lost 2 pounds). The data for the 12 subjects are:

Placebo	Treatment
2.0	-3.5
-3.1	-1.6
-1.0	-4.6
0.2	-0.9
0.3	-5.1
0.4	0.1

   Conduct a Bayesian analysis to compare the means of these two groups. Would you say the treatment is effective? Is your conclusion sensitive to the prior?

2. Load the classic Boston Housing Data in R:

   ```
   > library(MASS)
   > data(Boston)
   > ?Boston
   ```

   The response variable is medv, the median value of owner-occupied homes (in $1,000s), and the other 13 variables are covariates that describe the neighborhood.

   (a) Fit a Bayesian linear regression model with uninformative Gaussian priors for the regression coefficients. Verify the MCMC sampler has converged, and summarize the posterior distribution of all regression coefficients.

   (b) Perform a classic least squares analysis (e.g., using the lm function in R). Compare the results numerically and conceptually with the Bayesian results.

(c) Refit the Bayesian model with double exponential priors for the regression coefficients, and discuss how the results differ from the analysis with uninformative priors.

(d) Fit a Bayesian linear regression model in (a) using only the first 500 observations and compute the posterior predictive distribution for the final 6 observations. Plot the posterior predictive distribution versus the actual value for these 6 observations and comment on whether the predictions are reasonable.

3. Download the 2016 Presidential Election data from the book's website. Perform Bayesian linear regression with the response variable for county i being the difference between the percentage of the vote for the Republican candidate in 2016 minus 2012 and all variables in the object $\mathbf{X}$ as covariates.

(a) Fit a Bayesian linear regression model with uninformative Gaussian priors for the regression coefficients and summarize the posterior distribution of all regression coefficients.

(b) Compute the residuals $R_i = Y_i - \mathbf{X}_i\hat{\boldsymbol{\beta}}$ where $\hat{\boldsymbol{\beta}}$ is the posterior mean of the regression coefficients. Are the residuals Gaussian? Which counties have the largest and smallest residuals, and what might this say about these counties?

(c) Include a random effect for the state, that is, for a county in state $l = 1, ..., 50$,

$$Y_i|\alpha_l \sim \text{Normal}(\mathbf{X}_i\boldsymbol{\beta} + \alpha_l, \sigma^2)$$

where $\alpha_l \overset{iid}{\sim} \text{Normal}(0, \tau^2)$ and τ^2 has an uninformative prior. Why might adding random effects be necessary? How does adding random effects affect the posterior of the regression coefficients? Which states have the highest and lowest posterior mean random effect, and what might this imply about these states?

4. Download the US gun control data from the book's website. These data are taken from the cross-sectional study in [47]. For state i, let Y_i be the number of homicides and N_i be the population.

(a) Fit the model $Y_i|\boldsymbol{\beta} \sim \text{Poisson}(N_i\lambda_i)$ where $\log(\lambda_i) = \mathbf{X}_i\boldsymbol{\beta}$. Use uninformative priors and $p = 7$ covariates in $\mathbf{X}_i$: the intercept, the five confounders $\mathbf{Z}_i$, and the total number of gun laws in state i. Provide justification that the MCMC sampler has converged and sufficiently explored the posterior distribution and summarize the posterior of $\boldsymbol{\beta}$.

(b) Fit a negative binomial regression model and compare with the results from Poisson regression.

(c) For the Poisson model in (a), compute the posterior predictive distribution for each state with the number of gun laws set to zero. Repeat this with the number of gun laws set to 25 (the maximum number). According to these calculations, how would the number of deaths nationwide be affected by these policy changes? Do you trust these projections?

5. Download the titanic dataset from R,

```
library("titanic")
dat <- titanic_train
?titanic_train
```

Let $Y_i = 1$ if passenger i survived and $Y_i = 0$ otherwise. Perform a Bayesian logistic regression of the survival probability onto the passenger's age, gender (dummy variable) and class (two dummy variables). Summarize the effect of each covariate.

6. The T. rex growth chart data plotted in Figure 3.7 has $n = 6$ observations with weights (kg) 29.9, 1761, 1807, 2984, 3230, 5040, and 5654 and corresponding ages (years) 2, 15, 14, 16, 18, 22, and 28. Since weight must be positive, the gamma family of distributions is a reasonable model for these data. Describe a model with gamma likelihood and log mean that increases linearly with age. Approximate the posterior using MCMC, summarize the posterior distribution of all model parameters, and plot the data versus the fitted mean curve.

7. Consider the one-way random effects model $Y_{ij}|\alpha_i, \sigma^2 \sim$ Normal(α_i, σ^2) and $\alpha_i \sim$ Normal$(0, \tau^2)$ for $i = 1, ..., n$ and $j = 1, ..., m$. Assuming conjugate priors $\sigma^2, \tau^2 \sim$ InvGamma(a, b), derive the full conditional distributions of α_1, σ^2, and τ^2 and outline (but do not code) an MCMC algorithm to sample from the posterior.

8. Load the Gambia data in R:

```
> library(geoR)
> data(gambia)
> ?gambia
```

The response variable Y_i is the binary indicator that child i tested positive for malaria (pos) and the remaining seven variables are covariates.

(a) Fit the logistic regression model

$$\text{logit}[\text{Prob}(Y_i = 1)] = \sum_{j=1}^{p} X_{ij}\beta_j$$

with uninformative priors for the β_j. Verify that the MCMC sampler has converged and summarize the effects of the covariates.

(b) In this dataset, the 2,035 children reside in $L = 65$ unique locations (defined by the x and y coordinates in the dataset). Let $s_i \in \{1, ..., L\}$ be the label of the location for observation i. Fit the random effects logistic regression model

$$\text{logit}[\text{Prob}(Y_i = 1)] = \sum_{j=1}^{p} X_{ij}\beta_j + \alpha_{s_i} \text{ where } \alpha_l \overset{iid}{\sim} \text{Normal}(0, \tau^2)$$

and the β_j and τ^2 have uninformative priors. Verify that the MCMC sampler has converged; explain why random effects might be needed here; discuss and explain any differences in the posteriors of the regression coefficients that occur when random effects are added to the model; plot the posterior means of the α_l by their spatial locations and suggest how this map might be useful to malaria researchers.

9. Download the `babynames` data in R and compute the log odds of a baby being named "Sophia" each year after 1950:

```
library(babynames)
dat <- babynames
dat <- dat[dat$name=="Sophia" &
           dat$sex=="F" &
           dat$year>1950,]
yr  <- dat$year
p   <- dat$prop
t   <- dat$year - 1950
Y   <- log(p/(1-p))
```

Let Y_t denote the sample log-odds in year $t+1950$. Fit the following time series (auto-regressive order 1) model to these data:

$$Y_t = \mu_t + \rho(Y_{t-1} - \mu_{t-1}) + \varepsilon_t$$

where $\mu_t = \alpha + \beta t$ and $\varepsilon_t \overset{iid}{\sim} \text{Normal}(0, \sigma^2)$. The priors are $\alpha, \beta \sim \text{Normal}(0, 100^2)$, $\rho \sim \text{Uniform}(-1, 1)$, and $\sigma^2 \sim \text{InvGamma}(0.1, 0.1)$.

(a) Give an interpretation of each of the four model parameters: α, β, ρ, and σ^2.

(b) Fit the model using JAGS for $t > 1$, verify convergence, and report the posterior mean and 95% interval for each parameter.

(c) Plot the posterior predictive distribution for Y_t in the year 2020.

10. Open and plot the galaxies data in R using the code below,

```
> library(MASS)
> data(galaxies)
> ?galaxies
> Y <- galaxies
> hist(Y,breaks=25)
```

Model the observations $Y_1, ..., Y_{82}$ using a mixture of $K = 3$ normal distributions. For each the S MCMC iterations evaluate the density function on the grid $y \in \{5000, 5100, ..., 40000\}$ (351 points in the grid), giving an $S \times 351$ matrix of posterior samples. Plot the posterior median and 95% credible set of the density function at each of the 351 grid values. Does this mixture model fit the data well?

5

Model selection and diagnostics

CONTENTS

A statistical model is mathematical representation of the system that includes errors and biases in the observation process, and therefore lays bare the assumptions being made in the analysis. Of course, no statistical model is absolutely correct; in reality, functional relationships are not linear, errors are not exactly Gaussian, residuals are not independent, etc. Nonetheless, slight deviations in fit are a small price to pay for a simple and interpretable representation of reality that allows us to probe for important relationships, test scientific hypotheses and make predictions. On the other hand, if a model's assumptions are blatantly violated then the results of the analysis cannot be taken seriously. Therefore, selecting an appropriate model that is as simple as possible while fitting the data reasonably well is a key step in any parametric statistical analysis.

In this chapter we discuss Bayesian model selection and goodness-of-fit measures. For model selection, we assume there is a finite collection of candidate models denoted as $\mathcal{M}_1, ..., \mathcal{M}_M$. For example, we might compare Gaussian and student-t models,

$$\mathcal{M}_1 : Y_i \overset{iid}{\sim} \text{Normal}(\mu, \sigma^2) \text{ versus } \mathcal{M}_2 : Y_i \overset{iid}{\sim} \text{t}_\nu(\mu, \sigma^2) \qquad (5.1)$$

or whether or not to include a covariate

$$\mathcal{M}_1 : Y_i \overset{indep}{\sim} \text{Normal}(\beta_1, \sigma^2) \text{ versus } \mathcal{M}_2 : Y_i \overset{indep}{\sim} \text{Normal}(\beta_1 + X_i\beta_2, \sigma^2). \qquad (5.2)$$

The methods we discuss are general and can be used for other model-selection tasks, including to select random effects structure, the link function, etc.

Sections 5.1–5.5 introduce techniques for comparing and selecting statistical models. Section 5.1 begins with cross validation, which is a flexible and intuitive way to compare methods. For a more formal Bayesian treatment of model selection and assessment of model uncertainty, Section 5.2 introduces the Bayes factor and Section 5.3 provides computational tools to approximate Bayes factors when many models are under consideration. An attractive feature of Bayes factors is that rather than selecting a single model, uncertainty about the model is captured using posterior probabilities. Section 5.4 uses these posterior probabilities to make predictions that appropriately average over model uncertainty. While Bayes factors are appealing, they typically require extensive derivation or computation and are sensitive to the prior and so Section 5.5 provides less formal but more broadly applicable model selection criteria. In virtually any analysis none of the M models will be "right," and we are simply searching for the one that fits the "best." Therefore, Section 5.6 provides goodness-of-fit tools to verify that a selected model captures the important features of a dataset.

5.1   Cross validation

Arguably the most intuitive way to compare models is based on out-of-sample prediction performance. Ideally an independent validation set is available and used to evaluate performance. For example, for data streaming in over time one might train the M models on data available at the time of the analysis and use data collected after the analysis to measure predictive performance. Often this is not feasible as data are not collected sequentially, and so internal cross validation (CV) is used instead. In K-fold CV, each observation is randomly assigned to one of the K folds, with $g_i \in \{1, ..., K\}$ denoting the group assignment for observation i. Denote the subset of the data in fold k as $\mathbf{Y}_k = \{Y_i; g_i = k\}$ and the data in all other folds as $\mathbf{Y}_{(k)} = \{Y_i; g_i \neq k\}$. The model is then fit K times, with the kth model fit using $\mathbf{Y}_{(k)}$ to train the model and make predictions for $\mathbf{Y}_k$. In this way, a prediction is made for each observation using an analysis that excludes the observation, approximating out of sample prediction performance.

Bayesian prediction is based on the posterior predictive distribution (PPD) of the test set observations $\mathbf{Y}_k$ given the training data $\mathbf{Y}_{(k)}$ (Section 1.5). An advantage of this method of prediction is that it naturally averages over uncertainty in the model parameters. Generating samples from the PPD averaging over parameter uncertainty is straightforward using MCMC. Let $Y_i^{(s)}$ be the prediction made at MCMC iteration s for test set observation Y_i, then $Y_i^{(1)}, ..., Y_i^{(S)}$ can be used to approximate the PPD. For example, we might compute the Monte Carlo sample mean $\hat{Y}_i = \sum_{s=1}^{S} Y_i^{(s)}/S$, median $\tilde{Y}_i$, and τ-quantile $q_i(\tau)$ to approximate the posterior predictive mean, median and quantiles respectively.

Many metrics are available to summarize the accuracy of the predictions from each model [38]. The most common measures of point prediction are bias, mean squared error and mean absolute deviation,

$$BIAS = \frac{1}{n}\sum_{i=1}^{n}(\hat{Y}_i - Y_i)$$

$$MSE = \frac{1}{n}\sum_{i=1}^{n}(\hat{Y}_i - Y_i)^2$$

$$MAD = \frac{1}{n}\sum_{i=1}^{n}|\tilde{Y}_i - Y_i|,$$

with MSE being more sensitive to large errors than MAD. Performance of credible intervals can be summarized using empirical coverage and average width of prediction intervals

$$COV = \frac{1}{n}\sum_{i=1}^{n}I\left[q_i(\alpha/2) \le Y_i \le q_i(1-\alpha/2)\right]$$

$$WIDTH = \frac{1}{n}\sum_{i=1}^{n}\left[q_i(1-\alpha/2) - q_i(\alpha/2)\right],$$

where $I(A) = 1$ if the statement A is true and zero otherwise. A measure of fit to the entire distribution is the log score,

$$LS = \frac{1}{n}\sum_{i=1}^{n}\log[f(Y_i|\hat{\boldsymbol{\theta}}_i)] \tag{5.3}$$

where $\hat{\boldsymbol{\theta}}_i$ is the parameter estimate based on the fold that excludes observation i. The log score is the average log likelihood of the test set observations given the parameter estimates from the training data. Based on these measures, we might discard models with COV far below the nominal $1 - \alpha$ level and from the remaining model choose the one with small MSE, MAD and $WIDTH$ and large LS. It is essential that the evaluation is based on out-of-sample predictions rather than within-sample fit. Overly complicated models (e.g., a linear model with too many predictors) may replicate the data used to fit the model but be too unstable to predict well under new conditions.

Cross validation can be motivated by information theory. Suppose the "true" data generating model has PDF f_0 so that in reality $Y_i \overset{iid}{\sim} f_0$. Of course, we cannot know the true model and so we choose between M models with PDFs $f_1, ..., f_M$. Our objective is to select the model that is in some sense the closest to the true model. A reasonable measure of the difference between the true and postulated model is the Kullback–Leibler divergence

$$KL(f_0, f_j) = \mathrm{E}\left\{\log\left[\frac{f_j(Y)}{f_0(Y)}\right]\right\} = \mathrm{E}\left\{\log[f_j(Y)]\right\} - \mathrm{E}\left\{\log[f_0(Y)]\right\},$$

where the expectation is with respect to the true model $Y \sim f_0$. The term $\mathrm{E}\{\log[f_0(Y)]\}$ is the same for all $j = 1, ..., M$ and therefore ranking models based on $KL(f_0, f_j)$ is equivalent to ranking models based their log score $LS_j = \mathrm{E}\{\log[f_j(Y)]\}$. Since the data are generated as from f_0, the cross validation log score in (5.3) is a Monte Carlo estimate of the true log score LS_j, and therefore ranking models based on their cross validation log score is an attempt to rank them based on similarity to the true data-generating model.

5.2 Hypothesis testing and Bayes factors

Bayes factors [48] provide a formal summary of the evidence that the data support one model over another. For now, say there are only $M = 2$ models under consideration. Although not necessary for the computation of Bayes factors, most model comparison problems can be framed so that both models are nested in a common model and distinguished by the model parameters. For example, the full model might be $Y_i|\mu \overset{iid}{\sim} \text{normal}(\mu, \sigma^2)$, with model $\mathcal{M}_1$ defined by $\mu \leq 0$ and model $\mathcal{M}_2$ defined by $\mu > 0$. In a Bayesian analysis, the unknown parameters are treated as random variables. If the statistical models are stated as functions of the parameters, then the models are also random variables. For example, the posterior probability $\text{Prob}(\mu \leq 0|\mathbf{Y})$ is the posterior probability of model $\mathcal{M}_1$ and $\text{Prob}(\mu > 0|\mathbf{Y}) = \text{Prob}(\mathcal{M}_2|\mathbf{Y})$. These posterior probabilities are the most intuitive summaries of model uncertainty.

Posterior model probabilities incorporate information from both the data and the prior. Bayes factors remove the effect of the prior and quantify the data's support of the models. The Bayes factor (BF) of Model 2 relative to Model 1 is the ratio of posterior odds to prior odds,

$$BF = \frac{\text{Prob}(\mathcal{M}_2|\mathbf{Y})/\text{Prob}(\mathcal{M}_1|\mathbf{Y})}{\text{Prob}(\mathcal{M}_2)/\text{Prob}(\mathcal{M}_1)}. \tag{5.4}$$

If the prior probability of the two models are equal, then the BF is simply the posterior odds $\text{Prob}(\mathcal{M}_2|\mathbf{Y})/\text{Prob}(\mathcal{M}_1|\mathbf{Y})$; on the other hand, if the data are not at all informative about the models and the prior and posterior odds are the same, then $BF = 1$ regardless of the prior.

Selecting between two competing models is often referred to as hypothesis testing. In hypothesis testing one of the models is referred to as the null model or null hypothesis and the other is the alternative model/hypothesis. Hypothesis tests are usually designed to be conservative so that the null model is rejected in favor of the alternative only if the data strongly support this model. If we define Model 1 as the null hypothesis and Model 2 as an alternative hypothesis, then a rule of thumb [48] is that $BF > 10$ provides strong evidence of the alternative hypothesis (Model 2) compared to the null hypothesis (Model 1) and $BF > 100$ is decisive evidence.

A common mistake in a frequentist hypothesis testing is to state the probability that the null hypothesis is true; from a frequentist perspective, the parameters and thus the hypotheses are fixed quantities and not random variables, and therefore it is not sensible to assign them probabilities. From the Bayesian perspective however giving the posterior probability of each model/hypothesis is a legitimate summary of uncertainty.

Computing BFs for problems with many parameters can be challenging. In general, model selection can be framed as treating the model as an unknown random variable $\mathcal{M} \in \{\mathcal{M}_1, \mathcal{M}_2\}$ with prior probability $\text{Prob}(\mathcal{M} = \mathcal{M}_j) = q_j$. Conditioned on model j, i.e., $\mathcal{M} = \mathcal{M}_j$, the remainder of the Bayesian model is

$$\mathbf{Y} \sim f(\mathbf{Y}|\boldsymbol{\theta}; \mathcal{M}_j) \text{ and } \boldsymbol{\theta} \sim \pi(\boldsymbol{\theta}|\mathcal{M}_j). \tag{5.5}$$

Therefore, the two models can have different likelihood and prior functions. The BF requires the marginal posterior of the model integrating over uncertainty in the parameters,

$$\text{Prob}(\mathcal{M} = \mathcal{M}_j|\mathbf{Y}) = \int p(\boldsymbol{\theta}, \mathcal{M} = \mathcal{M}_j|\mathbf{Y})d\boldsymbol{\theta}. \tag{5.6}$$

Unfortunately, the marginalizing over the parameters (as in m in Table 1.4) is rarely possible. Also, the marginal distribution of $\mathcal{M}$ is not defined for improper priors. Therefore, *BFs cannot be used with improper priors*. Even with proper priors, BFs can very be sensitive to the choice of hyperparameters as shown by the examples below.

MCMC provides a means of approximating BFs in special cases. The BF for nested models defined by intervals of the parameters is straightforward to compute using MCMC. For example, in the model above with $Y_i|\mu \overset{iid}{\sim}$ normal(μ, σ^2) and $\mu \sim \text{Normal}(0, 10^2)$, and models $\mathcal{M}_1$ defined by $\mu \leq 0$ and $\mathcal{M}_2$ defined by $\mu > 0$, the posterior probability of Model 1 (2) can be approximated by generating samples from the posterior of the full model and recording the proportion of the samples for which μ is negative (positive). Section 5.3 provides a more general computational strategy for computing model probabilities.

A final note on Bayes factors is that they resemble the likelihood ratio statistic, which is commonly used in frequentist hypothesis testing. Assuming equal priors, the Bayes factor is

$$BF = \frac{\int f(\mathbf{Y}|\boldsymbol{\theta}; \mathcal{M}_2)\pi(\boldsymbol{\theta}|\mathcal{M}_2)d\boldsymbol{\theta}}{\int f(\mathbf{Y}|\boldsymbol{\theta}; \mathcal{M}_1)\pi(\boldsymbol{\theta}|\mathcal{M}_1)d\boldsymbol{\theta}}, \tag{5.7}$$

i.e., the ratio of marginal distribution of the data under the two models. This resembles the likelihood ratio from classical hypothesis testing

$$LR = \frac{f(\mathbf{Y}|\hat{\boldsymbol{\theta}}_2)}{f(\mathbf{Y}|\hat{\boldsymbol{\theta}}_1)}, \tag{5.8}$$

where $\hat{\boldsymbol{\theta}}_j$ is the MLE under model $j = 1, 2$. Therefore, both measures compare models based on the ratio of their likelihood, but the BF integrates over posterior uncertainty in the parameters whereas the likelihood ratio statistic plugs in point estimates.

Beta-binomial example: To build intuition about BFs, we begin with the univariate binomial model and test whether the success probability equals 0.5. Let $Y|\theta \sim \text{Binomial}(n, \theta)$ and consider two models for θ:

$$\mathcal{M}_1 : \theta = 0.5 \text{ versus } \mathcal{M}_2 : \theta \neq 0.5 \text{ and } \theta \sim \text{Beta}(a, b). \tag{5.9}$$

The first model has no unknown parameters, and the second model's parameter θ can be integrated out giving the beta-binomial model for Y (see Appendix A.1). Therefore, these hypotheses about θ correspond to two different models for the data

$$\mathcal{M}_1 : Y \sim \text{Binomial}(n, 0.5) \text{ versus } \mathcal{M}_2 : Y \sim \text{BetaBinom}(n, a, b). \tag{5.10}$$

Assuming priors $\text{Prob}(\mathcal{M}_j) = q_j$ and observed data $Y = y$, the BF of Model 2 relative to Model 1 is

$$BF(y) = \frac{f_{BB}(y; n, a, b)}{f_B(y; n, 0.5)} \tag{5.11}$$

where f_{BB} and f_B are the beta-binomial and binomial PMFs, respectively.

Figure 5.1 plots $BF(y)$ for $n = 20$ and different hyperparameters a and b. Assuming the uniform prior ($a = b = 1$) or Jeffreys' prior ($a = b = 0.5$) the BF exceeds 10 for $y < 5$ or $y > 15$ successes, and exceeds 100 for $y < 4$ or $y > 16$ successes; these scenarios give strong and decisive evidence, respectively, against the null hypothesis that $\theta = 0.5$ in favor of the alternative that $\theta \neq 0.5$. The BF is less than 10 for all possible y for the strong prior centered on 0.5 ($a = b = 50$). Under this prior the two hypotheses are similar and it is difficult to distinguish between them. Finally, under the strong prior that θ is close to one ($a = 50$ and $b = 1$), the BF exceeds 10 only for $y > 16$ successes; in this case y near zero is even less likely under the alternative than under the null that $\theta = 0.5$.

Normal-mean example: Say there is a single observation $Y|\mu \sim \text{Normal}(\mu, 1)$ and the objective is to test whether $\mu = 0$. The competing hypotheses are

$$\mathcal{M}_1 : \mu = 0 \text{ versus } \mathcal{M}_2 : \mu \neq 0 \text{ and } \mu \sim \text{Normal}(0, \tau^2). \tag{5.12}$$

Given that we observe $Y = y$, it can be shown that the BF of $\mathcal{M}_2$ relative to $\mathcal{M}_1$ is

$$BF(y) = \left(1 + \tau^2\right)^{-1/2} \exp\left[\frac{y^2}{2}\left(\frac{\tau^2}{1 + \tau^2}\right)\right]. \tag{5.13}$$

For fixed τ, $BF(y)$ increases to infinity as y^2 increases as expected because data far from zero contradict $\mathcal{M}_1 : \mu = 0$. However, for any fixed y, $BF(y)$

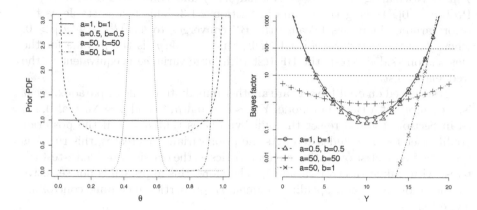

FIGURE 5.1
Bayes factor for the beta-binomial model. (left) Beta(a, b) prior PDF
for several combinations of a and b and (right) observed data Y versus the
Bayes factor comparing the beta-binomial model $Y|\theta \sim$ Binomial(n, θ) and
$\theta \sim$ Beta(a, b) versus the null model $Y \sim$ Binomial$(n, 0.5)$ for $n = 20$ and
several combinations of a and b.

converges to zero as the prior variance τ^2 increases. Therefore, even if the
observation is 10 standard deviation units above zero, the BF favors the null
model with $\mu = 0$ if the prior variance is sufficiently large. This is an example
of Lindley's paradox [51] where Bayesian tests can perform poorly depending
on the prior. For the normal means problem, this odd result occurs because
with large τ these data are unlikely under both models. For example, the
probability of $|Y| \in [9, 11]$ is very small under the null hypothesis that $\mu = 0$,
but under the alternative hypothesis this probability converges to zero as τ
increases because the marginal distribution of Y (averaging over μ) becomes
increasingly diffuse.

BFs are not defined with improper priors, and this example suggests that
they can have strange properties for uninformative priors. It has been argued
[9] that the standard half-Cauchy prior $\mu \sim t_1(0, 1)$ is preferred to the large-
variance normal prior because this prior is diffuse without having a variance
parameter to tune. In general, prior selection has more impact for hypothesis testing than estimation, making it imperative to report Bayes factors for
multiple priors to illuminate this sensitivity.

Lindley's paradox is more pronounced for tests of point null hypotheses
such as $\mathcal{M}_1 : \mu = 0$ than for one-sided tests such as

$$\mathcal{M}_1 : \mu \leq 0 \text{ versus } \mathcal{M}_2 : \mu > 0.$$

To compute the BF for these hypotheses we simply fit the Bayesian model

$Y|\mu \sim \text{Normal}(\mu, 1)$ and $\mu \sim \text{Normal}(0, \tau^2)$ and compute $\text{Prob}(\mathcal{M}_2|Y) = \text{Prob}(\mu > 0|Y)$ using the results in Section 2.1.3. This test is stable as the prior variance increases because the BF converges to $\Phi(y)/[1 - \Phi(y)] > 0$, where Φ is the standard normal CDF. Since $1 - \Phi(y)$ is the p-value for the classical one-sided z-test, the BF test with large variance is equivalent to the frequentist test.

Tests based on credible sets are another remedy to Lindley's paradox. That is, we simply fit a Bayesian model $Y|\mu \sim \text{Normal}(\mu, 1)$ and $\mu \sim \text{Normal}(0, \tau^2)$ as in Section 4.1 and reject the null hypothesis that $\mu = 0$ if the posterior credible set for μ excludes zero. As the prior variance increases, this rule has the same frequentist operating characteristics as the classic two-sided z-test (or t-test with unknown error variance). Therefore, this approach is not sensitive to the prior and has appealing frequentist properties including controlling Type I error.

5.3   Stochastic search variable selection

In Section 5.2, Bayes factors were used to summarize the data's support for $M = 2$ competing models. In many applications, the number of models under consideration is large and enumerating all models and computing each posterior probability is unfeasible. For example, in linear regression with p parameters there are $M = 2^p$ potential models formed by including subsets of the predictors; with $p = 30$ this is over a billion potential models. A classical way to overcome searching over all combinations of covariates is to employ a systematic search such as forward, backward or stepwise selection. In this section we discuss a stochastic alternative that randomly visits models according to their posterior probability.

Stochastic search variable selection (SSVS; [35]) approximates model probabilities using MCMC. SSVS introduces dummy variables to encode the model and then computes posterior probabilities of the dummy variables to approximate the posterior model probabilities. For example, consider the linear regression model

$$Y_i|\boldsymbol{\beta}, \sigma^2 \overset{indep}{\sim} \text{Normal}\left(\sum_{j=1}^p X_{ij}\beta_j, \sigma^2\right). \tag{5.14}$$

The $M = 2^p$ models formed by subsets of the predictors can be encoded by $\boldsymbol{\gamma} = (\gamma_1, ..., \gamma_p)$ where $\gamma_j = 1$ if covariate j is included in the model and $\gamma_j = 0$ if covariate j is excluded from the model, so that $\boldsymbol{\gamma} = c(1, 0, 1, 0, 0, ...)$ corresponds to the model $\text{E}(Y_i) = X_{i1}\beta_1 + X_{i3}\beta_3$.

A common prior over models is to fix $\gamma_1 = 1$ for the intercept and $\gamma_j|q \overset{iid}{\sim} \text{Bernoulli}(q)$ for $j > 1$, where q is the prior inclusion probabil-

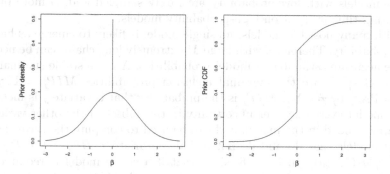

FIGURE 5.2
Spike and slab prior. PDF (left) and CDF of β under the spike and slab prior $\beta = \gamma\delta$ where $\gamma \sim \text{Bernoulli}(0.5)$ and $\delta \sim \text{Normal}(0,1)$.

ity. Given the model γ, the regression coefficients that are included in the model can have independent normal priors (other priors are possible, [70]) $\beta_j|\gamma_j = 1 \sim \text{Normal}(0, \sigma^2\tau^2)$. The inclusion probability and regression coefficient variance can be fixed or have prior such as $q \sim \text{Beta}(a, b)$ and $\tau^2 \sim \text{InvGamma}(\epsilon, \epsilon)$. As with Bayes factors (Section 5.2), the posterior for this model can be sensitive to the prior, and multiple priors should be compared to understand sensitivity.

The prior for β_j induced by this model is plotted in Figure 5.2. The prior is a mixture of two components: a peak at $\beta_j = 0$ corresponding to samples that exclude ($\gamma_j = \beta_j = 0$) covariate j and a Gaussian curve corresponding to samples that include ($\gamma_j = 1$ so $\beta_j = \delta_j$) β_j. Because of this distant shape, this prior is often called the spike-and-slab prior.

All M models can simultaneously be written as the supermodel

$$Y_i|\boldsymbol{\beta}, \sigma^2 \overset{indep}{\sim} \text{Normal}\left(\sum_{j=1}^{p} X_{ij}\beta_j, \sigma^2\right)$$
$$\beta_j = \gamma_j\delta_j$$
$$\gamma_j \sim \text{Bernoulli}(q)$$
$$\delta_j \sim \text{Normal}(0, \tau^2\sigma^2).$$

MCMC samples from this model include different subsets of the covariates. Posterior samples with $\gamma_j = 0$ have $\beta_j = 0$ and thus covariate j excluded from the model. This supermodel can be fit a single time and give approximations for all M posterior model probabilities. Since the search over models is done within MCMC, this is a stochastic search as opposed to systematic searches such as forward or backward regression. An advantage of stochastic search is that models with low probability are rarely sampled and so more of the computing time is spent on high-probability models.

With many possible models, no single model is likely to emerge as having high probability. Therefore, with large M extremely long chains can be needed to give accurate estimates of model probabilities. A more stable summary of the model space are the marginal inclusion probabilities $MIP_j = \mathrm{E}(\gamma_j = 1|\mathbf{Y}) = \mathrm{Prob}(\beta_j \neq 0|\mathbf{Y})$. MIP_j is the probability that covariate j is included in the model averaging over uncertainty in the subset of the other variables that are included in the model, and can be used to compute the Bayes factor for the models that do and do not include covariate j, $BF_j = [MIP_j/(1 - MIP_j)]/[q/(1 - q)]$. In fact, [6] show that if a single model is required for prediction, the model that includes covariates with MIP_j greater than 0.5 is preferred to the highest probability model.

Childhood malaria example: Diggle et al. [23] analyze data from $n = 1,332$ children from the Gambia. The binary response Y_i is the indictor that child i tested positive for malaria. We use five covariates in X_{ij}:

- Age: Age of the child, in days

- Net use: Indicator variable denoting whether (1) or not (0) the child regularly sleeps under a bed-net

- Treated: Indicator variable denoting whether (1) or not (0) the bed-net is treated (coded 0 if netuse=0)

- Green: Satellite-derived measure of the greenness of vegetation in the immediate vicinity of the village (arbitrary units)

- PCH: Indicator variable denoting the presence (1) or absence (0) of a health center in the village

All five covariates are standardized to have mean zero and variance one. We use the logit regression model

$$\mathrm{logit}[\mathrm{Prob}(Y_i = 1)] = \alpha + \sum_{j=1}^{p} X_{ij}\beta_j. \tag{5.15}$$

The spike-and-slab prior for β_j is $\beta_j = \gamma_j\delta_j$ where $\gamma_j \sim \mathrm{Bernoulli}(0.5)$ and $\delta_j \sim \mathrm{Normal}(0,\tau^2)$. Listing 5.1 gives JAGS code to fit this model.

Table 5.1 gives the model probabilities, i.e., the proportion of the MCMC samples with γ corresponding to each model. Only three models have posterior probability greater than 0.01. All three models include age, net use and greenness, and differ based on whether they include bed-net treatment, the health center indicator, or both. The marginal inclusion probabilities MIP_j in Table 5.2 exceed 0.5 for all covariates and therefore the best single model for prediction likely includes all covariates [6]. The posterior density for the regression coefficient corresponding to age (Figure 5.3) is bell-shaped because

Listing 5.1
JAGS code for SSVS.

```
1   for(i in 1:n){
2     Y[i]          ~ dbern(pi[i])
3     logit(pi[i]) <- alpha            + X[i,1]*beta[1] +
4                     X[i,2]*beta[2] + X[i,3]*beta[3] +
5                     X[i,4]*beta[4] + X[i,5]*beta[5]
6   }
7   for(j in 1:5){
8     beta[j] <- gamma[j]*delta[j]
9     gamma[j] ~ dbern(0.5)
10    delta[j] ~ dnorm(0,tau)
11  }
12  alpha ~ dnorm(0,0.01)
13  tau   ~ dgamma(0.1,0.1)
```

TABLE 5.1
Posterior model probabilities for the Gambia analysis. All other models have posterior probability less than 0.01.

Covariates	Probability
Age, Net use, Greenness, Treated	0.42
Age, Net use, Greenness, Treated, Health center	0.37
Age, Net use, Greenness, Health center	0.20

TABLE 5.2
Marginal posteriors for the Gambia analysis. Posterior inclusion probabilities (i.e., $\text{Prob}(\beta_j \neq 0|\mathbf{Y})$) and posterior median and 90% intervals for the β_j.

Covariate	Inclusion Prob	Median	95% Interval
Age	1.00	0.26	(0.19, 0.34)
Net use	1.00	-0.25	(-0.34, -0.17)
Greenness	1.00	0.29	(0.21, 0.37)
Treated	0.79	-0.13	(-0.24, 0.00)
Health center	0.56	-0.05	(-0.19, 0.00)

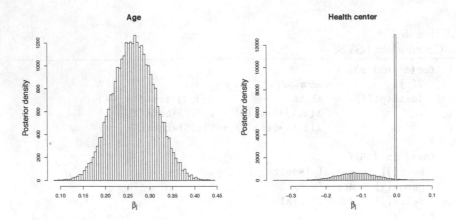

FIGURE 5.3
Posterior distribution for the SSVS analysis. Posterior distribution for the regression coefficients β_j for age and proximity to a health center.

it escapes the prior spike at zero, however the posterior for proximity to a health center retains considerable mass at zero and thus the spike-and-slab shape.

High-dimensional regression example: The data for this example are from [50] and can be downloaded from http://www.ncbi.nlm.nih.gov/geo (accession number GSE3330). In this study, $n = 60$ mice (31 female) were sampled and the physiological phenotype stearoyl-CoA desaturase 1 (SCD1) is taken as the response to be regressed onto the expression levels of 22,575 genes. Following [12] and [84], we use only the $p = 1,000$ genes with highest pairwise correlation with the response as predictors in the model. Even after this simplification, this leaves a high-dimensional problem with $p > n$.

We use the linear regression model with SSVS prior

$$Y_i \sim \text{Normal}\left(\alpha + \sum_{j=1}^{p} X_{ij}\beta_j, \sigma_e^2\right) \qquad (5.16)$$

$$\beta_j = \gamma_j \delta_j$$
$$\gamma_j \sim \text{Bernoulli}(q)$$
$$\delta_j \sim \text{Normal}(0, \sigma_e^2 \sigma_b^2).$$

Because p is large, it is possible to learn about the hyperparameters q and σ_b^2. We select priors $q \sim \text{Beta}(1,1)$ and $\sigma_e^2, \sigma_b^2 \sim \text{InvGamma}(0.1, 0.1)$ (Prior 1) and present the marginal inclusion probabilities $\text{Prob}(\beta_j \neq 0|Y)$. In high-dimensional problems, sensitivity to the prior is especially concerning. Therefore, we also refit with priors $q \sim \text{Beta}(1,2)$ (Prior 2) and $\sigma_b^2 \sim$

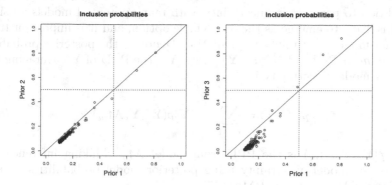

FIGURE 5.4

High-dimensional regression example. Marginal inclusion probabilities ($MIP_j = \text{Prob}(\beta_j \neq 0|\mathbf{Y})$) under three priors: (1) $q \sim \text{Beta}(1,1)$ and $\sigma_b^2 \sim$ InvGamma(0.1, 0.1), (2) $q \sim \text{Beta}(1,2)$ and $\sigma_b^2 \sim \text{InvGamma}(0.1, 0.1)$, and (3) $q \sim \text{Beta}(1,1)$ and $\sigma_b^2 \sim \text{InvGamma}(0.5, 0.5)$.

InvGamma(0.5, 0.5) (Prior 3). The models are fit using Gibbs sampling with 200,000 iterations, a burn-in of 20,000 iterations discarded and the remaining samples thinned by 20. This MCMC code requires 1–2 hours on an ordinary PC.

Only two genes have inclusion probability greater than 0.5 (Figure 5.4). The ordering of the marginal inclusion probabilities is fairly robust to the prior, but the absolute value of the marginal inclusion probabilities varies with the prior. The posterior median (95% interval) of the number of variables included in the model $p_{in} = \sum_{j=1}^{p} \gamma_j$ are 33 (12, 535) for Prior 1, 28 (12, 392) for Prior 2 and 20 (11, 64) for Prior 3. In this analysis, the results are more sensitive to the prior for the variance than the inclusion probability.

5.4 Bayesian model averaging

A main advantage of the Bayesian approach is the ability to properly handle uncertainty in model parameters when making statistical inference and predictions. In Section 1.5 (Figure 1.12) we explored the effect of accounting for parameter uncertainty in prediction. The posterior predictive distribution (PPD) for observation Y^* given the observed data $\mathbf{Y}$ averages over uncertainty in the parameters according to their posterior distribution,

$$p(Y^*|\mathbf{Y}) = \int p(Y^*, \boldsymbol{\theta}|\mathbf{Y})d\boldsymbol{\theta} = \int p(Y^*|\boldsymbol{\theta}, \mathbf{Y})p(\boldsymbol{\theta}|\mathbf{Y})d\boldsymbol{\theta}. \qquad (5.17)$$

In addition to parametric uncertainty, with many potential models no single model is likely to emerge as the only viable option, and it is important to account for model uncertainty in prediction. Denoting the posterior probability of model m as $\text{Prob}(\mathcal{M} = \mathcal{M}_m | \mathbf{Y}) = w_m(\mathbf{Y})$, the PPD of Y^* averaging over posterior model uncertainty is

$$p(Y^* | \mathbf{Y}) = \sum_{m=1}^{M} w_m(\mathbf{Y}) p(Y^* | \mathbf{Y}, \mathcal{M}_m) \qquad (5.18)$$

where $p(Y^* | \mathbf{Y}, \mathcal{M}_m)$ is the PPD from model $\mathcal{M}_m$. Making inference that accounts for model uncertainty using posterior model probabilities is called Bayesian model averaging (BMA; [45]).

SSVS (Section 5.3) via MCMC provides a convenient way to perform model-averaged predictions. Draws from the SSVS model include dummy indicators for different models as unknown parameters, and thus naturally average over models according to their posterior probability. Therefore, if a sample of Y^* is made at each MCMC iterations, then these samples follow the Bayesian model averaged PPD, as desired.

In addition to prediction, BMA can be used for inference on parameters common to all models. For example, if parameter β_j has posterior distribution $p(\beta_j | \mathbf{Y}, \mathcal{M}_m)$ under model $\mathcal{M}_m$, then the BMA posterior is $p(\beta_j | \mathbf{Y}) = \sum_{m=1}^{M} w_m(\mathbf{Y}) p(\beta_j | \mathbf{Y}, \mathcal{M}_m)$. However, BMA results for parameters must be carefully scrutinized because the interpretation of parameters can change considerably across models and so it is not always clear how to interpret an average over models. As an extreme case, in a regression analysis with collinearity the sign of β_j might change depending on the other covariates that are included, and so it is not obvious that results should be combined across models.

5.5   Model selection criteria

Cross validation and Bayes factors are both difficult to compute for large datasets and/or complicated models; cross validation requires fitting each model K times and Bayes factors require difficult integration. Model-fit criteria provide a useful alternative. The criteria considered in this chapter are defined via the deviance (twice the negative log likelihood) of the data given the parameters,

$$D(\mathbf{Y} | \boldsymbol{\theta}) = -2 \log[f(\mathbf{Y} | \boldsymbol{\theta})]. \qquad (5.19)$$

In this chapter, we compare only models with the same deviance function but different models/priors for the parameters so that the deviance is a comparable measure of fit across models. For some models, e.g., the random effect models in Section 4.4 which have equivalent representations conditional on and marginal over the random effects, the definition of the likelihood and thus deviance are not unique, but for most models for independent data the definition of the deviance is clear.

The Bayesian information criteria (BIC) is one such criteria, defined as

$$BIC = D(\mathbf{Y}|\hat{\boldsymbol{\theta}}_{MLE}) + \log(n)p \qquad (5.20)$$

where $\hat{\boldsymbol{\theta}}_{MLE}$ is the maximum likelihood estimate, n is the sample size and p is the number of parameters in the model. BIC is split into two terms: $D(\mathbf{Y}|\hat{\boldsymbol{\theta}}_{MLE})$ is small for models that fit the data well and $\log(n)p$ is small for simple low-dimensional models. Since simple models with good fit are desirable, models with smaller BIC are preferred.

BIC was originally motivated as an approximation to the Bayes factor. However, BIC has undesirable features from the Bayesian perspective. First, estimating $\boldsymbol{\theta}$ using the MLE does not use prior information and is not available from MCMC output. Second, quantifying model complexity with the number of parameter p does not account for informative priors. For example, if two models have the same number of parameters but one has very strongly informative priors and the other does not, then the model with informative priors should be regarded as simpler because it has fewer effective degrees of freedom.

The deviance information criteria (DIC; [77]) resolves these issues. From a Bayesian perspective, the deviance $D(\mathbf{Y}|\boldsymbol{\theta})$ is a random variable since is a function of the random variables $\boldsymbol{\theta}$, therefore $D(\mathbf{Y}|\boldsymbol{\theta})$ has a posterior distribution that can be summarized using its posterior mean $\bar{D}$. Model complexity is summarized by the effective number of parameters,

$$p_D = \bar{D} - \hat{D}, \qquad (5.21)$$

where $\hat{D} = D(\mathbf{Y}|\hat{\boldsymbol{\theta}})$ and $\hat{\boldsymbol{\theta}}$ is the posterior mean of $\boldsymbol{\theta}$. As shown by example below, p_D is typically not an integer because prior shrinkage can results in partial degrees of freedom. Both $\bar{D}$ and $\hat{\boldsymbol{\theta}}$ can be approximated using MCMC output by computing $D(\mathbf{Y}|\boldsymbol{\theta})$ and $\boldsymbol{\theta}$ at each iteration and taking the mean over iterations. Therefore, DIC is straightforward to compute given MCMC output.

The criteria is then

$$DIC = \bar{D} + p_D = D(\mathbf{Y}|\hat{\boldsymbol{\theta}}) + 2p_D. \qquad (5.22)$$

This resembles the Akaike information criterion (AIC) [2] $AIC = D(\mathbf{Y}|\hat{\boldsymbol{\theta}}_{MLE}) + 2p$ but with the posterior mean and effective degrees of freedom replacing the MLE and the total number of parameters. As with AIC and BIC, the actual value of the criteria is hard to interpret, but it can be used to rank models with models having small DIC being preferred (a loose rule of thumb is that a difference of 5 is substantial and 10 is definitive, but it is difficult to establish statistical significance using DIC). The intuition is that models with small DIC are simple (small p_D) and fit well (small $\bar{D}$). The effective number of parameters p_D is generally less than the number of parameters p if the prior are strong, as desired. Unfortunately p_D can be outside $[0, p]$ in pathological cases, typically where the posterior mean of $\boldsymbol{\theta}$ is not a good summary of the posterior as is the case in mixture models with multimodal priors and posteriors.

The motivation of p_D as a measure of model size is complex, but intuition can be built using a few examples. In multiple linear regression with $\mathbf{Y}|\boldsymbol{\beta} \sim \text{Normal}(\mathbf{X}\boldsymbol{\beta}, \sigma^2 \mathbf{I}_n)$ and Zellner's prior $\boldsymbol{\beta} \sim \text{Normal}(0, c\sigma^2(\mathbf{X}^T\mathbf{X})^{-1})$, the effective number of parameters is $p_D = \frac{c}{c+1}p$. In this case, the effective number of parameters increases from zero with tight prior ($c = 0$) to p with uninformative prior ($c = \infty$). Also, in the one-way random effects model (Section 4.4)

$$Y_{ij}|\mu_j \sim \text{Normal}(\mu_j, \sigma^2) \text{ and } \mu_j \sim \text{Normal}(0, c\sigma^2), \qquad (5.23)$$

for $i = 1, ..., n$ replications within each of the $j = 1, ..., p$ groups and fixed variance components σ^2 and c. The effective number of parameters is $p_d = \frac{c}{c+1/n}p$, which increases from 0 to p with the prior variance c.

The Watanabe–Akaike (also known as the widely applicable) information criteria ($WAIC$; [30]) is an alternative to DIC. $WAIC$ is proposed as an approximation to n-fold (i.e., leave-one-out) cross validation. Rather than the posterior mean of the deviance, $WAIC$ compute the posterior mean and variance of the likelihood and log likelihood. The criteria is

$$WAIC = -2 \sum_{i=1} \log\{\bar{f}_i\} + 2p_W \qquad (5.24)$$

where fit for observation i is measured by the posterior mean of the likelihood,

$$\bar{f}_i = \text{E}[f(Y_i|\boldsymbol{\theta})|\mathbf{Y}] \qquad (5.25)$$

and model complexity is measured by

$$p_W = \sum_{i=1}^{n} \text{Var}[\log(f(Y_i|\boldsymbol{\theta}))|\mathbf{Y}], \qquad (5.26)$$

defined as the sum of posterior variance of the log likelihood functions. Therefore, as with BIC and DIC, models with small $WAIC$ are preferred because they are simple and fit well.

Selecting a random effect model for the Gambia data: As in Section 5.3, the binary response Y_i indicates that child i tested positive for malaria and we consider covariates for age, bed-net use, bed-net treatment, greenness and health center. In this analysis we also consider child i's village $v_i \in \{1, ..., 65\}$ to account for dependence in the malaria status of children from the same village (Figure 5.5 plots the location and number of children sampled from each village). We use the random effects logistic regression model

$$\text{logit}[\text{Prob}(Y_i = 1)] = \alpha + \sum_{j=1}^{p} X_{ij}\beta_j + \theta_{v_i}, \qquad (5.27)$$

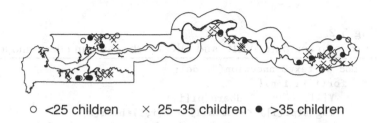

○ <25 children   × 25–35 children   ● >35 children

FIGURE 5.5
Gambia data. The location and number of children sampled from each village.

where θ_v is the random effect for village v. We compare three models for the village random effects via DIC and $WAIC$:

1. No random effects: $\theta_v = 0$
2. Gaussian random effects: $\theta_v \sim \text{Normal}(0, \tau^2)$
3. Double-exponential random effects: $\theta_v \sim \text{DE}(0, \tau^2)$

In all models, the priors are $\alpha, \beta_j \sim \text{Normal}(0, 100)$ and $\tau^2 \sim \text{InvGamma}(0.1, 0.1)$.

Code for Model 3 is given in Listing 5.2. DIC is computed using the `dic.samples` function in JAGS, although this unfortunately requires extra MCMC sampling. There is no analogous function for $WAIC$ in JAGS and so it must be computed outside of JAGS. In Listing 5.2 the extra line in the likelihood `like[i]` instructs JAGS to return posterior samples of the likelihood function $f(Y_i|\boldsymbol{\theta})$ which in this model is the binomial PMF (`dbin` in JAGS). After MCMC sampling, the posterior mean of `like[i]` is computed as the approximation to $\bar{f}_i$ and the posterior variance of `log(like[i])` is computed to approximate p_W.

The $WAIC$ and DIC results are in Table 5.3. Both measures show strong support for including village random effects, but cannot distinguish between Gaussian and double-exponential random-effect distributions. Since the Gaussian model is more familiar, this is probably the preferred model for these data.

2016 Presidential Election example: The data for this analysis come from Tony McGovern's very useful data repository [1]. The response variable, Y_i, is the percent increase in Republican (GOP) support from 2012 to 2016, i.e.,

$$100 \left(\frac{\% \text{ in } 2016}{\% \text{ in } 2012} - 1 \right), \tag{5.28}$$

---

[1]https://github.com/tonmcg/County_Level_Election_Results_12-16

Listing 5.2
JAGS code to compute WAIC and DIC for the random effects model.

```
1    mod <- textConnection("model{
2      for(i in 1:n){
3        Y[i]      ~ dbern(pi[i])
4        logit(pi[i]) <- beta[1] + X[i,1]*beta[2]
5                      + X[i,2]*beta[3] + X[i,3]*beta[4]
6                      + X[i,4]*beta[5] + X[i,5]*beta[6]
7                      + theta[village[i]]
8        like[i]      <- dbin(Y[i],pi[i],1) # For WAIC computation
9      }
10     for(j in 1:6){beta[j] ~ dnorm(0,0.01)}
11     for(j in 1:65){theta[j] ~ ddexp(0,tau)}
12     tau   ~ dgamma(0.1,0.1)
13   }")
14
15   data  <- list(Y=Y,X=X,n=n,village=village)
16   model <- jags.model(mod,data = data, n.chains=2,quiet=TRUE)
17   update(model, 10000, progress.bar="none")
18   samps <- coda.samples(model, variable.names=c("like"),
19                         n.iter=50000, progress.bar="none")
20
21   # Compute DIC
22   DIC   <- dic.samples(model,n.iter=50000,progress.bar="none")
23
24   # Compute WAIC
25   like  <- rbind(samps[[1]],samps[[2]]) # Combine the two chains
26   fbar  <- colMeans(like)
27   Pw    <- sum(apply(log(like),2,var))
28   WAIC  <- -2*sum(log(fbar))+2*Pw
```

TABLE 5.3
Model selection criteria for the Gambia data. *DIC* (p_D) and *WAIC* (p_W) for the three random effects models.

Random-effects model	DIC (p_D)	WAIC (p_W)
None	2526 (6.0)	2525 (6.0)
Gaussian	2333 (55.1)	2333 (53.4)
Double-exponential	2334 (57.1)	2333 (54.3)

in county $i = 1, ..., n$ (Figure 5.6a). The election data are matched with $p = 10$ county-level census variables (X_{ij}) obtained from Kaggle via Ben Hamner[2]:

1. Population, percent change – April 1, 2010 to July 1, 2014
2. Persons 65 years and over, percent, 2014
3. Black or African American alone, percent, 2014
4. Hispanic or Latino, percent, 2014
5. High school graduate or higher, percent of persons age 25+, 2009–2013
6. Bachelor's degree or higher, percent of persons age 25+, 2009–2013
7. Homeownership rate, 2009–2013
8. Median value of owner-occupied housing units, 2009–2013
9. Median household income, 2009–2013
10. Persons below poverty level, percent, 2009–2013.

All covariates are centered and scaled (e.g., Figure 5.6b). The objective is to determine the factors that are associated with an increase in GOP support. Also, following the adage that "all politics is local," we explore the possibility that the factors related to GOP support vary by state.

For a county in state s, we assume the linear model

$$Y_i = \beta_{0s} + \sum_{j=1}^{p} X_i \beta_{sj} + \varepsilon_i, \qquad (5.29)$$

where β_{js} is the effect of covariate j in state s and $\varepsilon_i \stackrel{iid}{\sim} \text{Normal}(0, \sigma^2)$. We compare three models for the β_{js}

1. Constant slopes: $\beta_{js} \equiv \beta_j$ for all counties
2. Varying slopes, uninformative prior: $\beta_{js} \stackrel{iid}{\sim} \text{Normal}(0, 10^2)$
3. Varying slopes, informative prior: $\beta_{js} \stackrel{indep}{\sim} \text{Normal}(\mu_j, \sigma_j^2)$

In all models, the prior for the error variance is $\sigma^2 \sim \text{InvGamma}(0.1, 0.1)$. In the first model the slopes have uninformative priors $\beta_j \sim \text{Normal}(0, 10^2)$. In the final model, the mean μ_j and variances σ_j^2 are given priors $\mu_j \sim \text{Normal}(0, 10^2)$ and $\sigma_j^2 \sim \text{InvGamma}(0.1, 0.1)$ and estimated from the data so that information is pooled across states via the prior. The three methods are compared using DIC and WAIC as in Listing 5.3 for the constant slopes model.

We first study the results from Model 1 with the same slopes in all states. Table 5.4 shows that all covariates other than home-ownership rate are associated with the election results. GOP support tended to increase in counties

---

[2]https://www.kaggle.com/benhamner/2016-us-election

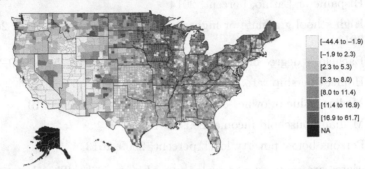

FIGURE 5.6

2016 Presidential Election data. Panel (a) plots the percentage change in Republican (GOP) support from 2012 to 2016 (Y_i) and Panel (b) plots the percent of counties with a bachelor's degree or higher (standardized to have mean zero and variance one; X_{i7}).

Listing 5.3
JAGS code to compute DIC for the constant slopes model.

```
1   model_string <- "model{
2     for(i in 1:n){
3       Y[i]    ~ dnorm(Xb[i],taue)
4       Xb[i]   ~ inprod(X[i,],beta[])
5       like[i] <- dnorm(Y[i],Xb[i],taue) # For WAIC
6     }
7     for(j in 1:p){beta[j] ~ dnorm(0,0.01)}
8     tau ~ dgamma(0.1,0.1)
9   }"
10
11  # Put the data and model statement JAGS format
12  dat   <- list(Y=Y,n=n,X=X,p=p)
13  init  <- list(beta=rep(0,p),tau=1)
14  model <- jags.model(textConnection(model_string),n.chains=2,
15                      inits=init,data = dat)
16
17  # Burn-in samples
18  update(model, 10000, progress.bar="none")
19
20  # Compute DIC
21  dic <- dic.samples(model1,n.iter=50000)
22
23  # Compute WAIC
24  samp <- coda.samples(model, variable.names=c("like"),
25               n.iter=50000)
26  like <- rbind(samps[[1]],samps[[2]]) # Combine the two chains
27  fbar <- colMeans(like)
28  Pw   <- sum(apply(log(like),2,var))
29  WAIC <- -2*sum(log(fbar)) + 2*Pw
```

TABLE 5.4
2016 Presidential Election multiple regression analysis. Posterior mean (95% interval) for the slopes β_j for the model with the same slopes in each state.

Covariate	Median	95% interval
Population change	-1.14	(-1.46, -0.81)
Percent over 65	0.93	(0.54, 1.32)
Percent African American	-1.56	(-1.89, -1.23)
Percent Hispanic	-2.06	(-2.40, -1.72)
Percent HS graduate	1.75	(1.25, 2.26)
Percent bachelor's degree	-6.19	(-6.71, -5.67)
Home-ownership rate	0.01	(-0.39, 0.41)
Median home value	-1.52	(-1.99, -1.06)
Median income	1.88	(1.13, 2.61)
Percent below poverty	1.48	(0.91, 2.04)

with decreasing population, high proportion of seniors and high school graduates, low proportions of African Americans and Hispanics, high income but low home value and high poverty rate.

The DIC ($\bar{D}$, p_D) for the three models are 21312 (21300, 12) for Model 1 with constant slopes, 18939 (18483, 455) for Model 2 with varying slope and uninformative priors, and 18842 (18604, 238) for Model 3 with varying slopes and informative priors. The first model is not rich enough to capture the important trends in data and thus has high $\bar{D}$ and DIC. The second model has the best fit to the observed data (smallest $\bar{D}$), but is too complicated (large p_D). The final model balances model complexity and fit and has the smallest DIC. WAIC gives similar results, with $WAIC$ (p_W) equal to 21335 (20), 18971 (406) and 18909 (259) for Models 1–3, respectively.

Inspection of the posterior of the variances σ_j^2 from the Model 3 shows that the covariate effect that varies the most across states is the proportion of the county with a Bachelor's degree. Figure 5.7 maps the posterior mean and standard deviation of the state-level slopes, β_{s7}. The association between the proportion of the population with a Bachelor's degree and change in GOP support is the strongest (most negative) in New England and the Midwest. The posterior standard deviation is the smallest in Colorado and Texas, possibly because these states have high variation in the covariate across counties.

Simulation study to evaluate the performance of DIC and $WAIC$: In these examples DIC and $WAIC$ gave similar results. However, in practice there will obviously be cases where they differ and the user will have to choose between them and defend their choice. Also, when applied to real data as above the "correct" model is unknown and so we cannot say for certain that either method selected the right model. One way to build trust in these criteria (or any other statistical method) is to evaluate their performance for simulated

(a) Effect of college graduates – posterior mean

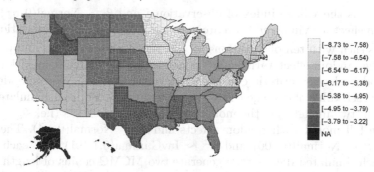

(b) Effect of college graduates – posterior SD

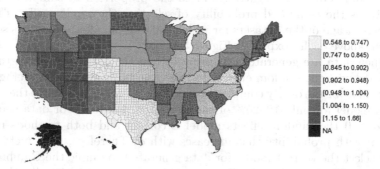

FIGURE 5.7

Results of the 2016 Presidential Election analysis. Posterior mean and standard deviation of the effect of the bachelor-degree rate on GOP support (β_{s7}) for the model with a different slope in each state and informative prior.

data where the correct model is known (see Section 7.3 for more discussion of simulation studies).

The data for this simulation experiment are generated to mimic the Gambia data. The response Y_i is binary and generated from the random effects logistic regression model

$$\text{logit}[\text{Prob}(Y_i = 1)] = \alpha + X_i\beta + \theta_{v_i},$$

where v_i is the village index of observation i and $\theta_v \sim \text{Normal}(0, \sigma^2)$ is the random effect for village v. Data are generated with $n = 100$ observations, ten villages each with ten observations, $\alpha = 0$, $\beta = 1$ and $X_i \stackrel{iid}{\sim} \text{Normal}(0,1)$. We vary the random effect variance σ^2 to determine how large it must be before the model selection criteria consistently favor the random effects model.

For each value of σ we generate $N = 100$ datasets. For each simulated data set we fit two models: (1) the model without random effects (i.e., $\theta_v = 0$) and (2) the full model with random effects and $\theta_v \sim \text{Normal}(0, \sigma^2)$. The priors are $\alpha, \beta \sim \text{Normal}(0, 100)$ and $\sigma^2 \sim \text{InvGamma}(0.1, 0.1)$. For each model and each simulated data set we generate two MCMC chains of length 10,000 (after a burn-in of 1,000) iterations and compute both DIC and $WAIC$. Say DIC_{mj} and $WAIC_{mj}$ are the criteria for model $m \in \{1,2\}$ for simulated dataset $j \in \{1, ..., N\}$. Figure 5.8 plots the N samples of $DIC_{2j} - DIC_{1j}$ (left) and $WAIC_{2j} - WAIC_{1j}$ (right). The random effects model is selected if these difference are negative, and so the proportion of samples that are negative is the estimated probability of selecting the random effects model. The percentage of the datasets for which the random effects model is selected is given above each boxplot in Figure 5.8.

When data are generated with $\sigma = 0$ then the correct model is the simple model without random effects. In this case, both metrics on average have larger value for the overly complex random effects model and select the random effects model for only 10–20% of the simulated datasets. For data generated with $\sigma > 0$ the random effects model is correct and both methods reliably select it with probability that increases with σ. Therefore, both methods reliably select the correct model for data generated to mimic the Gambia data. For this particular setting DIC returns the correct model with slightly higher probability than $WAIC$, but of course we cannot generalize this result to other settings.

5.6 Goodness-of-fit checks

Thus far we have discussed methods to choose from a prespecified set of models. This model selection step is crucial but cannot guarantee that the selected model actually fits the data well. For example, if all models considered are

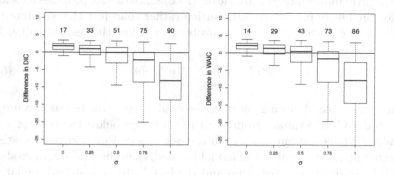

FIGURE 5.8
Simulation to evaluate selection criteria. Boxplots of the difference
in *DIC* (left) and *WAIC* (right) comparing random effects logistic regression with simple logistic regression for $N = 100$ simulated datasets
generated with random effect standard deviation σ. Each boxplot represents the distribution of the difference over N datasets simulated with $\sigma \in$
$\{0.00, 0.25, 0.50, 0.75, 1.00\}$ and the numbers above the boxplots are the percentage of the N datasets for which the difference was negative and thus the
criteria favored the random effects model.

Gaussian but the data are not, then even the best fitting model is inappropriate. Therefore, in addition to comparing models, diagnostics should be
performed to determine if the models capture the important features of the
data.

Standard diagnostic tools are equally important to a Bayesian and non-
Bayesian analysis. For example, in a linear regression, normality (e.g., residual
qq-plot), linearity (e.g., added variable plots), influential points (e.g., Cook's
D), etc., should be scrutinized. Many of these classic tools are based on least-
squares residuals and therefore are not purely Bayesian, but they remain valuable informal goodness-of-fit measures.

Another way to critique a model is out-of-sample prediction performance.
Say the data are split into a training set $\mathbf{Y}$ and a test set $\mathbf{Y}^* = (Y_1^*, ..., Y_m^*)$.
Section 1.5 discusses posterior predictive distribution (PPD) of the test set
observation i given the training data (and averaging over uncertainty in model
parameters), $f_i^*(y|\mathbf{Y})$. Comparing the test-set data to the PPD is a way to
verify the model fits well. The PPD evaluated at the observed test observation,

$$CPO_i = f_i^*(Y_i^*|\mathbf{Y}), \qquad (5.30)$$

is called the conditional predictive ordinate (CPO) [27, 65]. Test set observations with small CPO do not fit the model well and are potentially outliers.

A more interpretable diagnostic is to check that roughly 95% of the test-
set observations fall in the 95% posterior prediction intervals. For continuous

data, the probability integral transform (PIT) statistic [20] provides a measure of fit for the entire predictive distribution rather than just the 95% intervals. The PIT is the posterior predictive probability below the test set value, Y_i^*,

$$PIT_i = \int_{-\infty}^{Y_i^*} f_i^*(y|\mathbf{Y})dy. \tag{5.31}$$

This integral looks daunting, but can be easily approximated as the proportion of the MCMC samples from the PPD that are below the test-set value. Typically PIT_i is computed for each test-set observation and these statistics are plotted in a histogram. If the model fits well, then the PIT statistics should follow a Uniform(0,1) distribution and the PIT histogram should be flat.

An important consideration when interpreting diagnostic measures is that even a model that appears to be perfectly calibrated (say with uniform PIT statistics) is not necessarily the true model (if there is such a thing). For example, say the data are generated as $Y|X \sim$ Normal$(X,1)$ with $X \sim$ Normal$(0,1)$, then the model $Y \sim$ Normal$(0,2)$ will fit the data perfectly well, but is clearly inferior to a model that includes X as a predictor. Therefore, both model selection and goodness-of-fit testing are important.

Posterior predictive checks: Rather than focusing on predicting individual test set observations, posterior predictive checks (e.g., [33]) evaluate fit using summaries of the dataset. Let $\tilde{\boldsymbol{\theta}}$ be a posterior sample of the model parameters and $\tilde{\mathbf{Y}}$ be a replicate dataset drawn from the model given $\tilde{\boldsymbol{\theta}}$. To facilitate comparisons, the replicate dataset should have the same dimensions as the observed data. For example, in the linear regression model $\mathbf{Y} \sim$ Normal$(\mathbf{X}\boldsymbol{\beta}, \sigma^2\mathbf{I})$, the parameters are $\tilde{\boldsymbol{\theta}} = (\tilde{\boldsymbol{\beta}}, \tilde{\sigma}^2)$ and we would sample

$$\tilde{\mathbf{Y}}|\tilde{\boldsymbol{\theta}} \sim \text{Normal}(\mathbf{X}\tilde{\boldsymbol{\beta}}, \tilde{\sigma}^2\mathbf{I}) \tag{5.32}$$

using the same covariates $\mathbf{X}$ as in the original dataset.

Assume that MCMC produces S posterior samples $\boldsymbol{\theta}^{(1)}, ..., \boldsymbol{\theta}^{(S)}$ and we generate S replicate datasets $\tilde{\mathbf{Y}}_1, ..., \tilde{\mathbf{Y}}_S$ where $\tilde{\mathbf{Y}}_s|\boldsymbol{\theta}^{(s)} \sim p(\mathbf{y}|\boldsymbol{\theta}^{(s)})$. If the model fits well, then $\mathbf{Y}$ and the $\tilde{\mathbf{Y}}_s$ should follow the same distribution, and thus comparing $\mathbf{Y}$ to the distribution of $\tilde{\mathbf{Y}}_s$ provides an evaluation of whether the proposed data-generating model is valid. Summarizing fit for an entire multivariate distribution such as the distribution of $\mathbf{Y}$ is challenging, and so we restrict comparisons to one-number summaries of the data set, $D(\mathbf{Y})$. For example, $D(\mathbf{Y})$ could be the mean or maximum value of the dataset, or any other measure of interest. A visual goodness-of-fit check plots the predictive distribution of the summary statistics, $D(\tilde{\mathbf{Y}}_1), ..., D(\tilde{\mathbf{Y}}_S)$, as a histogram and compares the observed measure $D(\mathbf{Y})$ to this distribution; if the observed value falls far from the center of the distribution then this is evidence the model does not fit well.

Selecting the summary measure D is clearly important. The most effective summary measures are those that verify modeling assumptions. For example, when fitting the model $Y \sim$ Normal(μ, σ^2), the mean and variance of the data should fall in the predictive distribution of the mean and variance because there are parameters in the model for these summaries. Therefore, selecting D to the sample mean or variance is not informative. However, the normal model assumes the distribution is symmetric and there are no parameters in the model to capture asymmetry. Therefore, taking D to be the skewness provides a useful verification that this modeling assumption holds.

The Bayesian p-value is a more formal summary of the posterior predictive check. The Bayesian p-value is the probability under repeated sampling from the fitted model of observing a summary statistic at least as large as the observed statistic. This probability can be approximated using MCMC output as the proportion of the S draws $D(\tilde{\mathbf{Y}}_1), ..., D(\tilde{\mathbf{Y}}_S)$ that are greater than $D(\mathbf{Y})$. A Bayesian p-value near zero or one indicates that in at least one aspect the model does not fit well.

The Bayesian p-value resembles the p-value from classical hypothesis testing in that both quantify the probability under repeated sampling of observing a statistic at least as large as the observed statistic. An important difference is that the classical p-value assumes repeated sampling under the null hypothesis, whereas the Bayesian p-value assumes repeated sampling from the fitted Bayesian model. Another important difference is that for the Bayesian p-value, probabilities near either zero or one provide evidence against the fitted model.

Gun control example: Kalesan et al. (2016) [47] study the relationship between state gun laws and firearm-related death rates. The response variable, Y_i, is the number of firearm-related deaths in 2010 in state i. The analysis includes five potential confounders (Z_{ij}): 2009 firearm death rate per 10,000 people; firearm ownership rate quartile; unemployment rate quartile; non-firearm homicide rate quartile and firearm export rate quartile. The covariates of interest in the study are status of gun laws in the state. Let X_{il} indicate that state i has law l. In this example, we simply use the number of laws $X_i = \sum_l X_{il}$ as the covariate. Setting aside correlation versus causation issues, the objective of the analysis is to determine if there is a relationship between the number of gun laws in the state and its firearm-related death rate. Our objective in this section is to illustrate the use of posterior predictive checks to verify that the model fits well.

We check the fit of two models. The first is the usual Poisson regression model

$$Y_i \sim \text{Poisson}(\lambda_i) \text{ where } \lambda_i = N_i \exp\left(\alpha + \sum_{j=1}^{5} Z_{ij}\beta_j + X_i\beta_6\right) \quad (5.33)$$

and N_i is the state's population. A concern with the Poisson model is that because the mean equals the variance it may not be flexible enough to capture large counts. Therefore, we also consider the negative binomial model with mean λ_i and over-dispersion parameter m

$$Y_i \sim \text{NB}\left(\frac{m}{\lambda_i + m}, m\right). \quad (5.34)$$

Listing 5.4
JAGS code for the over-dispersed Poisson regression model.

```
1    # Likelihood
2    for(i in 1:n){
3      Y[i]            ~ dnegbin(q[i],m)
4      q[i]            <- m/(m+N[i]*lambda[i])
5      log(lambda[i]) <- alpha + inprod(Z[i,],beta[1:5]) +
                          X[i]*beta[6]
6    }
7
8    #Priors
9    for(j in 1:6){
10       beta[j] ~ dnorm(0,0.1)
11   }
12   alpha ~ dnorm(0,0.1)
13   m     ~ dgamma(0.1,0.1)
14
15   # Posterior predictive checks
16   for(i in 1:n){
17     Y2[i]   ~ dnegbin(q[i],m)
18     rate[i] <- Y2[i]/N[i]
19   }
20
21   D[1] <- min(Yp[])
22   D[2] <- max(Yp[])
23   D[3] <- max(Yp[])-min(Yp[])
24   D[4] <- min(rate[])
25   D[5] <- max(rate[])
26   D[6] <- max(rate[])-min(rate[])
```

The priors for the fixed effects are $\alpha, \beta_j \sim \text{Normal}(0, 10)$ and for the negative-binomial model $m \sim \text{Gamma}(0.1, 0.1)$. Code to fit the negative-binomial model is given in Listing 5.4.

We interrogate the models using posterior predictive checks with the following six test statistics:

1. Minimum count: $D_1(\mathbf{Y}) = \min\{Y_1, ..., Y_n\}$

2. Maximum count: $D_2(\mathbf{Y}) = \max\{Y_1, ..., Y_n\}$

3. Range of counts: $D_3(\mathbf{Y}) = \max\{Y_1, ..., Y_n\} - \min\{Y_1, ..., Y_n\}$

4. Minimum rate: $D_4(\mathbf{Y}) = \min\{Y_1/N_1, ..., Y_n/N_n\}$

5. Maximum rate: $D_5(\mathbf{Y}) = \max\{Y_1/N_1, ..., Y_n/N_n\}$

6. Range of rates: $D_6(\mathbf{Y}) = \max\{Y_1/N_1, ..., Y_n/N_n\} - \min\{Y_1/N_1, ..., Y_n/N_n\}$

We use test statistics related to the range of the counts and rates because

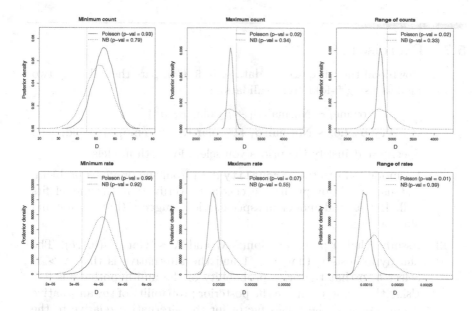

FIGURE 5.9
Bayesian p-values for the gun control example. The density curves
show the posterior predictive distribution of the six test statistics for the
Poisson and negative-binomial (NB) models, and the vertical lines are the
test statistics for the observed data. The Bayesian p-value is given in the
legend's parentheses.

the main concern here is properly accounting for large counts. Figure 5.9
plots the PPD for both models and all six test statistics. The PPD from the
Poisson model does not capture the largest counts or the range of counts observed in the dataset. The Bayesian p-value is close to zero or one for all four
test statistics involving either the maximum or the range. As expected, the
negative-binomial model gives much wider prediction intervals and thus less
extreme Bayesian p-values. Therefore, while the Poisson model is not appropriate for this analysis, the negative-binomial model appears to be adequate
based on these (limited) tests. However, in both analyses the coefficient associated with the number of gun laws (β_6) is negative with high probability
(95% interval (-0.017,-0.011) for the Poisson model and (-0.026, -0.008) for
the negative-binomial model), and so the conclusion that there is a negative
association between the number of gun laws in a state and its firearm-related
mortality is robust to this modeling assumption.

5.7   Exercises

1. Download the `airquality` dataset in R. Compare the following two models using 5-fold cross validation:

   $\mathcal{M}_1$: ozone$_i \sim$ Normal($\beta_1 + \beta_2$solar.R$_i, \sigma^2$)

   $\mathcal{M}_2$: ozone$_i \sim$ Normal($\beta_1 + \beta_2$solar.R$_i + \beta_2$temp$_i + \beta_3$wind$_i, \sigma^2$).

   Specify and justify the priors you select for both models.

2. Fit model $\mathcal{M}_2$ to the `airquality` data from the previous problem, and use posterior predictive checks to verify that the model fits well. If you find model misspecification, suggest (but do not fit) alternatives.

3. Assume that $Y|\lambda \sim$ Poisson($N\lambda$) and $\lambda \sim$ Gamma($0.1, b$). The null hypothesis is that $\lambda \leq 1$ and the alternative is that $\lambda > 1$. Select b so that the prior probability of the null hypothesis is 0.5. Using this prior, compute the posterior probability of the alternative hypothesis and the Bayes factor for the alternative relative to the null hypothesis for (a) $N = 10$ and $Y = 12$; (b) $N = 20$ and $Y = 24$; (c) $N = 50$ and $Y = 60$; (d) $N = 100$ and $Y = 120$. For which N and Y is there definitive evidence in favor of the alternative?

4. Use the "Mr. October" data (Section 2.4) $Y_1 = 563$, $N_1 = 2820$, $Y_2 = 10$, and $N_2 = 27$. Compare the two models:

   $\mathcal{M}_1$: $Y_1|\lambda_1 \sim$ Poisson($N_1\lambda_1$) and $Y_2|\lambda_2 \sim$ Poisson($N_2\lambda_2$)

   $\mathcal{M}_2$: $Y_1|\lambda_0 \sim$ Poisson($N_1\lambda_0$) and $Y_2|\lambda_0 \sim$ Poisson($N_2\lambda_0$).

   using Bayes factors, DIC and $WAIC$. Assume the Uniform($0,c$) prior for all λ_j and compare the results for $c = 1$ and $c = 10$.

5. Use DIC and $WAIC$ to compare logistic and probit links for the `gambia` data in the R package `geoR` using the five covariates in Listing 5.2 and no random effects.

6. Fit logistic regression model to the `gambia` data in the previous question and use posterior predictive checks to verify the model fits well. If you find model misspecification, suggest (but do not fit) alternatives.

7. For the NBA free throw data in Section 1.6, assume that for player i, $Y_i|p_i \sim$ Binomial(n_i, p_i) where Y_i is the number of clutch makes, n_i is the number of clutch attempts, and p_i is the clutch make probability. Compute the posterior probabilities of the models:

   $\mathcal{M}_1$: logit(p_i) $= \beta_1 +$ logit(q_i)

   $\mathcal{M}_2$: logit(p_i) $= \beta_1 + \beta_2$logit(q_i).

where q_i is the overall free throw percentage. Specify the priors you use for the models' parameters and discuss whether the results are sensitive to the prior.

8. Fit model $\mathcal{M}_2$ to the NBA data in the previous question and use posterior predictive checks to verify the model fits well. If you find model misspecification, suggest (but do not fit) alternatives.

9. Download the Boston Housing Data in R from the `Boston` dataset. The response is `medv`, the median value of owner-occupied homes, and the other 13 variables are covariates that describe the neighborhood. Use stochastic search variable selection (SSVS) to compute the most likely subset of the 13 covariates to include in the model and the marginal probability that each variable is included in the model. Clearly describe the model you fit including all prior distributions.

10. Download the `WWWusage` dataset in R. Using data from times $t = 5, ..., 100$ as outcomes (earlier times may be used as covariates), fit the autoregressive model

$$Y_t | Y_{t-1}, ..., Y_1 \sim \text{Normal}(\beta_0 + \beta_1 Y_{t-1} + ... + \beta_L Y_{t-L}, \sigma^2)$$

where Y_t is the WWW usage at time t. Compare the models with $L = 1, 2, 3, 4$ and select the best time lag L.

11. Using the `WWWusage` dataset in the previous problem, fit the model with $L = 2$ and use posterior predictive checks to verify that the model fits well. If you find model misspecification, suggest (but do not fit) alternatives.

12. Open and plot the galaxies data in R using the code below,

```
> library(MASS)
> data(galaxies)
> ?galaxies
> Y <- galaxies
> n <- length(Y)
> hist(Y,breaks=25)
```

Model the observations $Y_1, ..., Y_{82}$ using the Student-t distribution with location μ, scale σ and degrees of freedom k. Assume prior distributions $\mu \sim \text{Normal}(0, 10000^2)$, $1/\sigma^2 = \tau \sim \text{Gamma}(0.01, 0.01)$ and $k \sim \text{Uniform}(1, 30)$.

(a) Use posterior predictive checks to evaluate whether the t distribution captures the mean, variance, skewness and kurtosis of the data.

(b) Repeat this with k fixed at 30 (so that the model is essentially Gaussian).

6

Case studies using hierarchical modeling

CONTENTS

6.1 Overview of hierarchical modeling

Thus far we have introduced Bayesian ideas in the context of standard statistical models. However, one of the primary benefits of Bayesian methodology is flexibility to handle non-standard cases with irregularities such as missing values, censored data, variables measured with error, multiple data sources each with distinct biases and errors, subpopulations with different properties, etc. Incorporating all of these features in an analysis may seem daunting, but can often be accomplished by breaking the large model into manageable layers and combining the layers in a hierarchical model. Hierarchical modeling (also known as a multilevel model) is thus an essential model-building tool.

To see how building a model for complex data can be simplified by thinking hierarchically, say our objective is to specify the joint distribution of three variables X, Y and Z. Directly fitting a multivariate joint distribution can be challenging, especially if the three variables have different supports. However, any trivariate distribution can be written

$$f(x, y, z) = f(x)f(y|x)f(z|x, y). \qquad (6.1)$$

By ordering the three variables and specifying the univariate marginal distribution of X and then the univariate conditional distributions for $Y|X$ and $Z|X, Y$, the multivariate problem is reduced to three univariate problems. Because the variables are ordered and each conditional distribution depends only on the previous variables in the ordering, the resulting joint distribution is guaranteed to be valid. Also, since any multivariate distribution can be decomposed this way, there is no loss of flexibility by taking this approach.

The trivariate model is represented as a directed acyclic graph (DAG) in

(a)

(b)

FIGURE 6.1
Directed acyclic graphs (DAGs). Panel (a) shows the DAG for the model $f(X,Y,Z) = f(X)f(Y|X)f(Z|X,Y)$ and Panel (b) shows the DAG for the model $f(X,Y,Z) = f(X)f(Y|X)f(Z|Y)$.

Figure 6.1a. A DAG (also called a Bayesian network) represents the model as a graph with each observation and parameter as a node (i.e., the points that define the graph) and edges (i.e., connections between the nodes) to denote conditional dependence. To define a valid stochastic model the graph must be directed and acyclic. A directed graph associates each edge with a direction; an arrow from X to Y indicates that the hierarchical model is defined by modeling the conditional distribution of Y as a function of X. The absence of an arrow from X to Z in Figure 6.1b conveys the choice that conditioned on Y, Z does not depend on X, i.e., $f(z|x,y) = f(z|y)$; in contrast, Figure 6.1a is the DAG for the model with conditional dependence between X and Z given Y. The graph must also be acyclic, meaning that it is impossible to follow directed edges from a node through the graph and return to the original node. These two conditions rule out building models such as $p(x,y,z) = p(x|y,z)p(y|z)p(z|y)$, which may not be a valid joint distribution.

Hierarchical models can take many forms, but a general way to build a model is through a data layer, a process layer and a prior layer. Model building should begin with the *process layer* that contains the underlying scientific processes of interest and the unknown parameters. Building this layer is ideally done in consultation with domain experts. Once this layer is defined the statistical objectives can be articulated, for example, to estimate a particular parameter or test a specific hypothesis. Ideally these objectives dictate the data to be collected for the analysis. The *data layer* relates (via the likelihood function) the data to the process and encodes bias and error in the data collection procedure, which requires knowledge of how the data were collected. Finally the *prior layer* quantifies uncertainty about the model parameters at the onset of the analysis.

Building a model hierarchically is convenient, but not fundamentally different than models we have considered previously. In fact, we have already encountered many hierarchical models such as the random effects models in Section 4.4. This means that the computational methods to sample from the posterior described in Chapter 3 apply to hierarchical models as do the graphical and numerical methods used to summarize the posterior.

The hierarchical model for disease progression below is an example of a model built this way. Although we do not carry out an analysis, this example illustrates the model building process. Let S_t and I_t be the number of susceptible and infected individuals in a population, respectively, at time t. Scientific understanding of the disease is used to model disease propagation. In consultation with an epidemiologist, we might select the simple Reed–Frost model [1]:

$$\text{Process layer:} \quad I_{t+1} \sim \text{Binomial}\left[S_t, 1 - (1-q)^{I_t}\right]$$
$$S_{t+1} = S_t - I_{t+1}$$

where it is assumed that all infected individuals are removed from the population before the next time step and q is the probability of a non-infected person coming into contact with and contracting the disease from an infected individual. The epidemiological process-layer model expresses the disease dynamics up to a few unknown parameters. To estimate these parameters, the number of cases at time t, denoted Y_t, is collected. The data layer models our ability to measure the process I_t. For example, after discussing the data collection procedure with domain experts, we might assume there are no false positives (uninfected people counted as infected) but potentially false negatives (uncounted infected individuals) and thus

$$\text{Data layer: } Y_t|I_t \sim \text{Binomial}(I_t, p) \tag{6.2}$$

where p is the probability of detecting an infected individual. The Bayesian model is completed using priors

$$\text{Prior layer:} \quad I_1 \sim \text{Poisson}(\lambda_1), S_1 \sim \text{Poisson}(\lambda_2)$$
$$p, q \sim \text{beta}(a, b).$$

Figures 6.2 plots the DAG corresponding to this model.

Another general idea for building a hierarchical model is to split the data into homogeneous groups and then pool information across the groups via the prior distribution. The random slopes model for the jaw bone density data (plotted in Figure 4.6) discussed in Section 4.4 is an example of a hierarchical model built this way. The model is

$$\text{Data layer :} \quad Y_{ij}|\boldsymbol{\beta}_i \sim \text{Normal}(\mathbf{X}_j\boldsymbol{\beta}_i, \sigma^2)$$
$$\text{Process layer :} \quad \boldsymbol{\beta}_i|\boldsymbol{\mu}, \boldsymbol{\Sigma} \sim \text{Normal}(\boldsymbol{\mu}, \boldsymbol{\Sigma})$$
$$\text{Prior layer :} \quad \boldsymbol{\mu} \sim \text{Normal}(\mathbf{0}, c^2\mathbf{I}_2), \boldsymbol{\Sigma} \sim \text{InvWishart}(\nu, \boldsymbol{\Omega})$$

for $i = 1, ..., n = 20$ and $j = 1, ..., m = 4$. Each of the 20 patients gets its own simple linear regression model, and these regressions are combined in the process layer that specifies the distribution of the regression coefficients across individuals in the patient population.

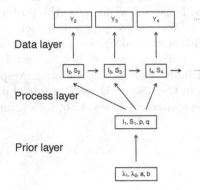

FIGURE 6.2
Directed acyclic graph for the Reed–Frost infectious disease model.

This model is visualized as a DAG in Figure 6.3. The DAG shows how information moves through the hierarchical model. For example, how does the data from patient 1 help us predict the bone density for patient 2's next visit? To traverse the DAG from $\mathbf{Y}_1$ to $\mathbf{Y}_2$ requires going through the population parameters $\boldsymbol{\mu}$ and $\boldsymbol{\Sigma}$. That is, $\mathbf{Y}_1$ informs the model about $\boldsymbol{\beta}_1$ which shapes the random effects distribution that enters the model for $\boldsymbol{\beta}_2$ and thus $\mathbf{Y}_2$. If we only had data from patient 2, then we would likely resort to an uninformative prior for $\boldsymbol{\beta}_2$ and with only a few observations for patient 2 the posterior would be unstable. However, the hierarchical model allows us to borrow strength across patients to stabilize the results.

MCMC is a natural choice for fitting hierarchical models; just as hierarchical models build complexity by layering simple conditional distributions, MCMC samples from the complex posterior distribution by sequentially updating parameters from simple full conditional distributions. In fact, displaying the hierarchical model as a DAG not only helps understand the model but it also aids in coding the MCMC sampler because the full conditional distribution for a parameter depends only on terms with an arrow to or from the parameter's node in the DAG. For example, from Figure 6.3 it is clear that the full conditional distribution for $\boldsymbol{\beta}_2$ depends only on the data-layer terms for $Y_{21}, ..., Y_{2m}|\boldsymbol{\beta}_2$ and the process-layer term for $\boldsymbol{\beta}_2|\boldsymbol{\mu}, \boldsymbol{\Sigma}$. If we view the model only through these terms then it is immediately clear that the full conditional distribution of $\boldsymbol{\beta}_2$ is exactly the full conditional distribution of the regression coefficients in standard Bayesian linear regression (Section 4.2).

When to stop adding layers? The hierarchical models in this section have three levels: data, process and prior. However, likely the values that define

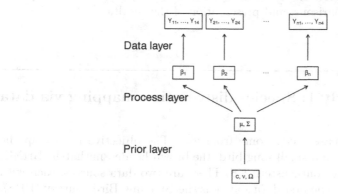

FIGURE 6.3
Directed acyclic graph. Visual representation of the random slopes model $Y_{ij}|\boldsymbol{\beta}_i \sim \text{Normal}(\mathbf{X}_j\boldsymbol{\beta}_i, \sigma^2)$ for $i = 1, ..., n$ and $j = 1, ..., 4$ with random effect distribution $\boldsymbol{\beta}_i|\boldsymbol{\mu}, \boldsymbol{\Sigma} \sim \text{Normal}(\boldsymbol{\mu}, \boldsymbol{\Sigma})$ and priors $\boldsymbol{\mu} \sim \text{Normal}(\mathbf{0}, c^2\mathbf{I}_2)$ and $\boldsymbol{\Sigma} \sim \text{InvWishart}(\nu, \boldsymbol{\Omega})$.

the prior layer will not be known exactly, and it is tempting to add a fourth (and a fifth, etc.) layer to explain this uncertainty. A general rule of thumb is to stop adding layers when there is no replication to estimate parameters. For example, referring to the DAG in Figure 6.3, it is reasonable to add a layer to estimate the random effect mean $\boldsymbol{\mu}$ and covariance $\boldsymbol{\Sigma}$ because there are repeated random effects $\boldsymbol{\beta}_1, ..., \boldsymbol{\beta}_n$ that can be leveraged to estimate these parameters. However, adding an additional layer to estimate prior mean $\boldsymbol{\Omega}$ of the random effect covariance $\boldsymbol{\Sigma}$ would not be reasonable because there is only one covariance $\boldsymbol{\Sigma}$ in the model, and even if we knew $\boldsymbol{\Sigma}$ exactly we would not be able to estimate its distribution from a single sample.

The remainder of this chapter is formatted as a sequence of case studies in hierarchical modeling. The three case studies each pose different challenges:

1. **Species distribution mapping via data fusion**: Combining information from multiple data streams while accounting for their bias and uncertainty

2. **Tyrannosaurid growth curves**: Pooling information across subpopulations (species) and quantifying uncertainty in non-linear models with small number of observations

3. **Marathon analysis with missing data**: Accounting for missing data when performing statistical inference and prediction

In these analyses we demonstrate the flexibility of hierarchical modeling, and also illustrate complete Bayesian analyses including model and prior specification, model comparisons, and presentation of the results.

6.2 Case study 1: Species distribution mapping via data fusion

The data for this case study come from [63]. The objective is to map the spatial distribution of a small song bird, the brown-headed nuthatch (BHNU; *Sitta pusilla*), in the southeastern US. There are two data sources, each with different strengths. The first data source is the Breeding Birds Survey (BBS). BBS is a network of hundreds of routes surveyed by thousands of volunteers that has been active since 1966 [74]. Data are collected systematically by trained volunteers and sites are visited annually to monitor for changes. However, even with this immense sampling effort, there are spatial and temporal gaps in BBS coverage. An emerging line of research is to supplement systematic survey data with massive citizen science data such as the Cornell Lab of Ornithology's eBird database [80], which consists of millions of data points from thousands of citizen scientists each year. These data are not collected by trained birders, but have far greater spatial and temporal coverage.

For this analysis, the southeast US is partitioned into $n = 741$, 0.25×0.25 degree lat/lon cells and we analyze data from 2012. Let N_{1i} be the number BBS sampling occasions in cell i and $Y_{1i} \in \{0, 1, ..., N_{1i}\}$ be the number of occasions with a sighting of the BHNU. Many cells do not have a BBS route and thus $N_{1i} = Y_{i1} = 0$. Similarly, for cell i define N_{2i} as the number of hours logged by eBird citizen scientists and Y_{2i} be the number of BHNU eBird sightings. Figure 6.4 maps the data. The BBS sampling effort is fairly uniform, whereas the eBird effort is more concentrated in populated areas. Both maps show more BHNU sightings in Alabama, Georgia and the Carolinas.

The true process of interest in cell i is the abundance $\lambda_i \geq 0$, defined as the expected number of birds present in the region during one unit of surveying effort. For a cell that is not inhabited by the BHNU $\lambda_i = 0$. The data layer relates the process to the observed data, and requires careful consideration of the strengths of the two data sources. First we make the assumption that the BBS and eBird datasets are independent given λ_i. This seems reasonable as most eBird users are not following BBS updates. The BBS data are gathered by expert birders and thus we assume that there are no false positives or negatives (although more flexible models are available and likely preferable, see [63] for example). If the number of birds present during a survey is distributed as Poisson(λ_i), then the probability that there is at least one bird present is $1 - \exp(-\lambda_i)$, and so we model the BBS data as $Y_{i1}|\lambda_i \sim \text{Binomial}[N_{1i}, 1 - \exp(-\lambda_i)]$.

For the eBird data, we do allow for false positives and false negatives. The

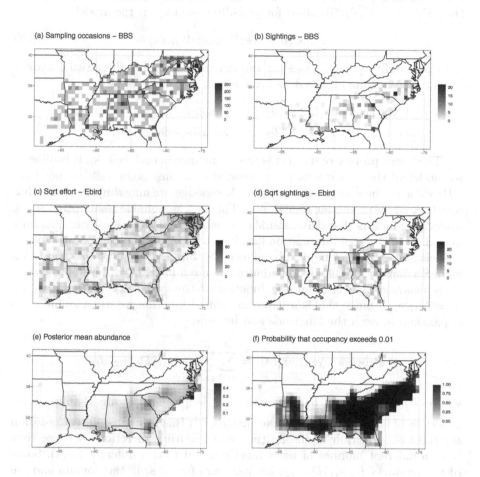

FIGURE 6.4
2012 Brown-headed nuthatch data. Panels (a) and (b) plot the number of BBS sampling occasions (N_{1i}) and number of BBS sightings (Y_{1i}); Panels (c) and (d) plot the square root of Ebird effort ($\sqrt{N_{2i}}$) and the square root of the number of Ebird sightings ($\sqrt{Y_{2i}}$). Panel (e) plots the posterior mean abundance, λ_i, and Panel (f) plots the posterior probability that the occupancy probability exceeds 0.01, i.e., $\text{Prob}[1 - \exp(-\lambda_i) > 0.01 | \mathbf{Y}]$.

mean is $N_{2i}\tilde{\lambda}_i$ with rate $\tilde{\lambda}_i = \theta_1\lambda_i + \theta_2$, where $\theta_1 > 0$ controls the difference between observation rates by BBS observers and eBird observers and $\theta_2 > 0$ is the eBird false positive rate so that if the cell is truly uninhabited and $\lambda_i = 0$, then $E(Y_{2i}) = N_{2i}\theta_2$. To allow for over-dispersion we fit the model

$$Y_{2i}|\lambda_i, \boldsymbol{\theta} \sim \text{NegBinomial}(q_i, m) \qquad (6.3)$$

with probability $q_i = m/(\tilde{\lambda}_i + m)$ and size m. Combining these two contributions to the likelihood, the data layer is

$$\text{Data layer} : Y_{1i}|\lambda_i, \theta_1, \theta_2 \sim \text{Binomial}[N_{1i}, 1 - \exp(-\lambda_i)]$$
$$Y_{2i}|\lambda_i, \theta_1, \theta_2 \sim \text{NegBinomial}(q_i, m).$$

The latent process of interest is the abundance in each cell, λ_i. It is difficult to model all the λ_i without prior knowledge because some cells do not have BBS data. In the absence of other prior knowledge, we may simply assume that nearby cells have similar abundance. This is a reasonable assumption if the underlying factors that drive abundance (climate, habitat, etc.) vary spatially and allows us to pool information locally to estimate abundance. Many spatial models can be used for model abundance [28], but here we use spline regression as in Section 4.5.1. The log abundance (we use a log transformation to ensure λ_i is non-negative) is a smooth function of the grid cell's spatial location, $\mathbf{s}_i = (s_{1i}, s_{2i})$. Since this is a two-dimensional function we use a spline basis expansions in both the longitude and latitude,

$$\text{Process layer} : \log(\lambda_i) = \sum_{j=1}^{J}\sum_{k=1}^{K} B_j(s_{i1})D_k(s_{i2})\beta_{jk} \qquad (6.4)$$

where B_j are B-spline basis functions of longitude, D_k are B-spline basis functions of latitude and $\beta_{jk} \overset{iid}{\sim} \text{Normal}(\beta_0, \sigma^2)$ (here we use different notation for the basis function in the latitude and longitude directions because they have a different number of basis functions and thus a different form). Some of the products $B_j(s_{i1})D_k(s_{i2})$ are near zero for all $\mathbf{s}_i$ in the domain and are discarded. Since the spatial domain spans a wider range of longitudes than latitudes, we take $K = 2L$ and $J = L$ for a total of $p = 2L^2$ terms of the form $X_l = B_j(s_{i1})D_k(s_{i2})$ and select L using DIC. To complete the Bayesian hierarchical model, we specify uninformative priors

$$\text{Prior layer} : \theta_1, \theta_2, m, \sigma^{-2} \sim \text{Gamma}(0.1, 0.1) \text{ and } \beta_0 \sim \text{Normal}(0, 100).$$
$$(6.5)$$

JAGS code to implement this model is given in Listing 6.1.

Two chains are run, each with a burn-in of 10,000 iterations and 50,000 post-burn-in samples. The samples are thinned by 5 leaving 20,000 samples to approximate the posterior. The DIC (p_D) is 3107 (30) for $L = 4$, 3056 (58) for $L = 6$, 3015 (89) for $L = 8$, 2999 (127) for $L = 10$, 3014 (177) for $L = 12$ and 3009 (209) for $L = 14$, and so we proceed with $L = 10$. With $L = 10$, the

Listing 6.1
Spatial data fusion model for BHNU abundance.

```
1    # Data layer
2      for(i in 1:n){
3        Y1[i]  ~ dbin(phi[i],N1[i]) # BBS
4        phi[i] <- 1-exp(-lam[i])
5
6        Y2[i]  ~ dnegbin(q[i],m)    # eBird
7        q[i]   <- m/(m+N2[i]*(theta1*lam[i]+theta2))
8      }
9
10   # Process layer
11     for(j in 1:p){beta[j]~dnorm(beta0,tau)}
12     for(i in 1:n){
13       log(lam[i]) <- inprod(X[i,],beta[])
14     }
15
16   # Prior layer
17     theta1 ~ dgamma(0.1,0.1)
18     theta2 ~ dgamma(0.1,0.1)
19     m      ~ dgamma(0.1,0.1)
20     tau    ~ dgamma(0.1,0.1)
21     beta0  ~ dnorm(0,1)
```

effective sample size is greater than 1,000 for all β_{jk}, indicating the sampler has mixed well and sufficiently explored the posterior.

Table 6.1 presents the posterior distributions of the hyperparameters. Of note, the eBird false positive rate, θ_2, is estimated to be near zero, and thus the eBird data appears to be a reliable source of information. The posterior mean of λ_i and the posterior probability that cells are occupied (i.e., at least one individual is present) are mapped in Figures 6.4e and 6.4f, respectively. As expected, the estimated abundance is the largest in Georgia and the Carolinas, but the occupancy probability is also high farther west in Louisiana and Arkansas. The occupancy probabilities would be lower in these western states if the eBird data were excluded.

6.3   Case study 2: Tyrannosaurid growth curves

We analyze the data from 20 fossils to estimate the growth curves of four tyrannosaurid species: Albertosaurus, Daspletosaurus, Gorgosaurus and Tyrannosaurus. The data are taken from Table 1 of [25] and plotted in Figure 6.5. The objective is to establish the growth curve, i.e., expected body mass by

TABLE 6.1
Posteriors for the BHNU analysis. Posterior median and 95% intervals mean for the final fit with $L = 10$.

	Median	95% Interval
Scaling factor, θ_1	11.5	(9.2, 14.4)
False positive rate, θ_2	0.00	(0.00, 0.00)
Over-dispersion parameter, m	0.45	(0.37, 0.55)
Mean abundance parameter, β_0	-5.81	(-6.69, -4.86)
Spline standard deviation, σ	5.58	(4.52, 7.03)

age, for each species. The data exhibit non-linear relationships between age and mass and there are commonalities between species. We therefore pursue a non-linear hierarchical model.

The original analysis of these data used non-linear least squares (fitted curves shown in the left panel of Figure 6.5). Quantifying uncertainty in this fit is challenging. The sampling distribution of the estimator does not have a closed form due to the non-linear mean structure, and with only a handful of observations to estimate roughly the same number of parameters, large-sample normal approximations are not valid and resampling techniques such as the bootstrap may have insufficient data to approximate the sampling distribution. As shown below, a Bayesian analysis powered by MCMC fully quantifies posterior uncertainty.

Let Y_{ij} and X_{ij} be the body mass and age, respectively, of sample i from species $j = 1, ..., 4$. We model the data as

$$Y_{ij} = f_j(X_{ij})\epsilon_{ij}, \tag{6.6}$$

where f_j is the true growth curve for species j and $\epsilon_{ij} > 0$ is multiplicative error with mean one. We use multiplicative error rather than additive error because variation in the population likely increases with mass/age. Assuming the errors are log-normal with $\log(\epsilon_{ij}) \sim \text{Normal}(-\sigma_j^2/2, \sigma_j^2)$ then $\text{E}(\epsilon_{ij}) = 1$ as required and the model becomes

$$\log(Y_{ij}) \sim \text{Normal}\left(\log[f_j(X_{ij})] - \sigma_j^2/2, \sigma_j^2\right), \tag{6.7}$$

and $\text{E}(Y_{ij}) = f_j(X_{ij})$ and where σ_j^2 controls the error variance for species j.

The data in Figure 6.5 (left) clearly exhibit nonlinearity. However, after taking a log transformation of both mass and age their relationship is fairly linear (Figure 6.5, right). Therefore, one model we consider is the log-linear model

$$\log[f_j(X)] = a_j + b_j \log(X) \tag{6.8}$$

where a_j and b_j are the intercept and slope, respectively, for species j. On the original scale, the corresponding growth curve is $f_j(X) = \exp(a_j)X^{b_j}$. If

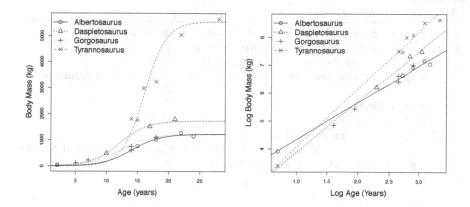

FIGURE 6.5
Tyrannosaurid growth curve data. Panel (a) gives scatter plots of the estimated age and body mass (kg) of 20 samples of four tyrannosaurid species; Panel (b) plots the same data after a log transformation of both variables. The curves plotted in Panel (a) are the fitted logistic curves from [25] and the lines in Panel (b) are least squares fits.

as expected b_j is positive, then the growth curve increases indefinitely, which may not be realistic. Therefore, we compare the log-linear model with the logistic growth curve

$$f_j(X) = a_j + b_j \frac{\exp[d_j(x - c_j)]}{1 + \exp[d_j(x - c_j)]}. \qquad (6.9)$$

where $x = \log(X)$. This model has four parameters:

1. a_j is the expected mass at age 0
2. b_j is the expected lifetime gain in mass
3. $\log(c_j)$ is the age at which the species reaches half its expected gain
4. $d_j > 0$ determines the rate of increase with age.

The form of the curve is increasing (assuming $b_j > 0$) and plateaus at $a_j + b_j$ rather than continuing to increase with age. This is the same function fit by [25], except that we transform to log age.

In addition to comparing these two forms of growth curves, we also compare two priors. The first prior ("unpooled") fits each species separately using uninformative priors. For the log-linear model the priors are $a_j, b_j \sim \text{Normal}(0, 10)$ and $\sigma_j^2 \sim \text{InvGamma}(0.1, 0.1)$ and for the logistic model the priors are $\log(a_j), \log(b_j), c_j, \log(d_j) \sim \text{Normal}(0, 10)$ and $\sigma_j^2 \sim$

Listing 6.2
JAGS code for hierarchical growth curve modeling.

```
1    # n is the total number of observations for all species
2    # x[i] is the log age of individual i
3    # y[i] is the log mass of individual i
4    # sp[i] is the species number (1, 2, 3, or 4) of individual i
5
6    # Data layer
7    for(i in 1:n){
8      y[i]      ~ dnorm(muY[i],taue)
9      muY[i]   <- log(a[sp[i]] + b[sp[i]]/(1+exp(-part[i]))) - 0.5/taue
10     part[i]  <- (x[i]-c[sp[i]])/d[sp[i]]
11   }
12
13   # Process layer
14   for(j in 1:N){
15     a[j]   <- exp(alpha[j,1])
16     b[j]   <- exp(alpha[j,2])
17     c[j]   <- alpha[j,3]
18     d[j]   <- exp(alpha[j,4])
19
20     for(k in 1:4){alpha[j,k] ~ dnorm(mu[k],tau[k])}
21   }
22
23   # Prior layer
24   for(k in 1:4){
25     mu[k]   ~ dnorm(0,0.1)
26     tau[k]  ~ dgamma(0.1,0.1)
27   }
28   taue ~ dgamma(0.1,0.1)
```

InvGamma$(0.1, 0.1)$. The normal prior is replaced with the log-normal prior for a_j, b_j and d_j to ensure these parameters are positive and thus $f_j(X)$ is positive and increasing for all X. The second prior ("pooled") is a Bayesian hierarchical model that borrows information across the four species. In the pooled analysis we assume the variance is the same for all species, $\sigma_j^2 = \sigma^2$ and has uninformative prior $\sigma^2 \sim$ InvGamma$(0.1, 0.1)$. For the log-linear model priors for the intercepts are $a_j \sim$ Normal(μ_a, σ_a^2), where $\mu_a \sim$ Normal$(0, 10)$ and $\sigma_a^2 \sim$ InvGamma$(0.1, 0.1)$. The same hierarchical model is applied to the $\log(a_j)$, $\log(b_j)$, c_j and $\log(d_j)$ in the logistic model. The JAGS code for this model is given in Listing 6.2.

This hierarchical model treats the parameters across the four species as random effects, and learning about the random effects distribution (i.e., μ_a and σ_a^2) stabilizes the posterior by providing additional information via the priors. It is debatable whether these parameters are truly random effects, i.e., whether there is an infinite distribution of exchangeable species from which these four

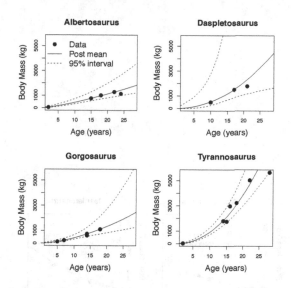

FIGURE 6.6
Fitted log-linear growth curves – unpooled. Observations (points) versus the posterior mean (solid lines) and 95% intervals (dashed lines) of the tyrannosaurid growth curves for the unpooled log-linear model.

were randomly selected for the study. However, analyzing the data from these four species using a random effects model clearly improves the results (as shown below) by pooling information across species to reduce uncertainty.

We fit the model with log-linear and logistic growth curves, each separate by species (unpooled) and using a hierarchical model (pooled). DIC (p_D) for the four fits are: 29 (25) for log-linear unpooled, -3 (9) for log-linear pooled, 64 (41) for logistic unpooled and -2 (12) for logistic pooled. The pooled models reduce model complexity (as measured by p_D) and this leads to smaller (better) DIC. DIC for the log-linear and logistic growth curves are similar.

Figures 6.6–6.9 plot the posterior mean and pointwise 95% credible interval for f_j for each model and each species (the interval estimates are for f_j and not Y_{ij}, so they should not include 95% of the observations). The posterior means of the four methods are fairly similar and all fit the data well. The main difference between the fits is that by borrowing information across species, the pooled analyses have narrower credible sets. Visually, the log-linear fits in Figure 6.9 appear to sufficiently model the growth curves. However, given that the logistic curve fits nearly as well and possesses the intuitive property of plateauing at an advanced age, this model is arguably preferable when considering the entire life course.

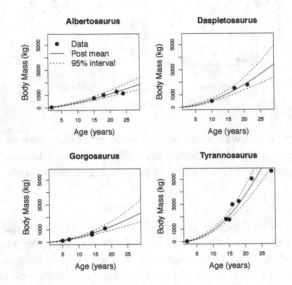

FIGURE 6.7
Fitted log-linear growth curves – pooled. Observations (points) versus
the posterior mean (solid lines) and 95% intervals (dashed lines) of the tyrannosaurid growth curves for the pooled (hierarchical) log-linear model.

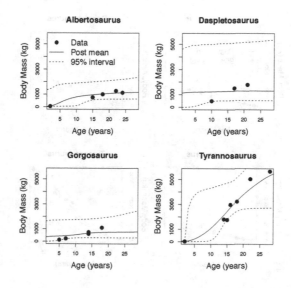

FIGURE 6.8
Fitted log-linear growth curves – logistic, unpooled. Observations
(points) versus the posterior mean (solid lines) and 95% intervals (dashed
lines) of the tyrannosaurid growth curves for the unpooled logistic model.

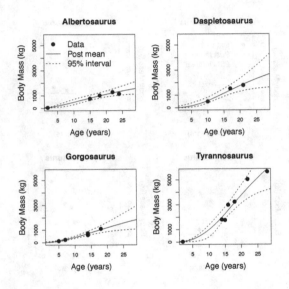

FIGURE 6.9
Fitted log-linear growth curves – logistic, pooled. Observations (points)
versus the posterior mean (solid lines) and 95% intervals (dashed lines) of the
tyrannosaurid growth curves for the pooled (hierarchical) logistic model.

6.4   Case study 3: Marathon analysis with missing data

Missing data are a complicating factor in many analyses. An an example, consider the 2016 Boston Marathon data from Section 2.1.7. In this analysis, we build a linear regression model to predict the speed (minutes per mile) of the final mile (mile 26) of the top female runners as a function of their speeds in the first 25 miles. Let Y_i be the speed of mile 26 for runner $i = 1, ..., n = 149$ and X_{ij} be the speed for runner i in mile $j = 1, ..., p = 25$. Approximately 8% of the speed measurements are missing: the number of missing observations ranges from 0 to 19 across runners and the percentage of missing observations ranges from 0% to 55% (miles 6 and 7) across miles (Figure 6.10a).

The easiest resolution to the missing-data problem is to simply discard observations with missing data and proceed with a complete-case linear regression. Discarding observations with at least one missing X_{ij} would reduce the sample size from 149 runners to 58. Given that most of the missing values are for miles 6 and 7 and these covariates are likely not important predictors of the response, it would be wasteful to discard all of these observations from the analysis. Another simple approach is to impute the missing X_{ij} using the sample mean of the X_{ij} for the other runner's speed at mile j or a linear regression using the other covariates in the model. A drawback of these single-imputation approaches is that they do not account for uncertainty in the missing observations, and thus the resulting posterior inference is questionable. Multiple imputation techniques are also available [72] (and are often motivated using Bayesian ideas).

A Bayesian analysis using a hierarchical model is a natural way to handle missing data. The Bayesian approach handles missing data in the same way as unknown parameters; we represent our uncertainty about them by treating them as random variables in a hierarchical Bayesian model. As with an unknown parameter, posterior inference on an unknown missing covariate requires assigning it a prior distribution. A fairly general model is

$$Y_i | \mathbf{X}_i \sim \text{Normal}(\mathbf{X}_i \boldsymbol{\beta}, \sigma^2) \text{ where } \mathbf{X}_i \sim \text{Normal}(\boldsymbol{\mu}, \boldsymbol{\Sigma}) \qquad (6.10)$$

where $\boldsymbol{\mu}$ is the mean vector of length p and $\boldsymbol{\Sigma}$ is the $p \times p$ covariance matrix of the covariates. In the absence of subject-matter prior information, the hyperparameters $\boldsymbol{\mu}$ and $\boldsymbol{\Sigma}$ are given priors and estimated as part of the Bayesian analysis, resembling a random-effects model. In this way, the complete cases inform the model about the distribution of the covariates across observations (via $\boldsymbol{\mu}$ and $\boldsymbol{\Sigma}$), and this information is used to impute the missing values.

Of course, this approach requires a reasonable model for the covariate distribution. This is challenging when p is large and/or the covariates are non-Gaussian. For example, if the covariates are a mix of continuous and binary variables, sufficiently capturing their joint distribution is difficult. A simple approach is to model the p covariates with independent priors. This is inefficient but at least supplies a reasonable approximation in many situations.

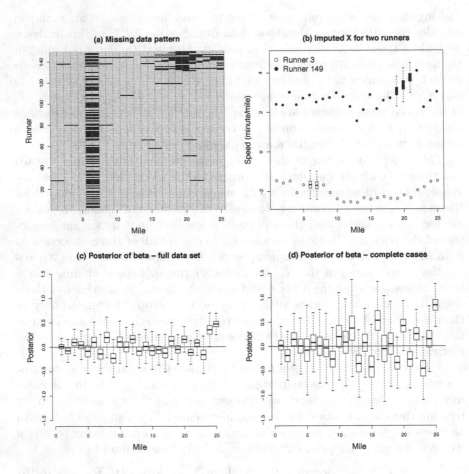

FIGURE 6.10
Missing data analysis of the 2016 Boston Marathon data. Panel (a) shows the missing (black) and non-missing (white) X_{ij} by runner (i) and mile (j). Panel (b) plots the observed (points) standardized covariates (X_{ij}) and the posterior distributions (boxplots) of the missing covariates for two runners. Panels (c) and (d) plot the posterior distribution of each regression coefficient, β_j, for the missing-data model and complete-case analysis, respectively.

More sophisticated modeling, such as the methods outlined in Section 4.5, may improve this aspect of the analysis. A crucial (and often unverifiable) assumption is that there are no systematic biases in the missing data, i.e., they are missing completely at random. For the present example, a hypothetical source of systematic bias is that missing times are caused by runners moving too fast to record their speeds. If this is the case, it would be impossible to observe this bias because the data are missing, and the model-based imputations would underestimate the missing speeds leading to questionable inference.

Assuming that the covariate model is correct, a hierarchical Bayesian analysis appropriately accounts for uncertainty in the missing values. A Gibbs sampler would update each parameter (β, σ, etc.) and then cycle through the missing observations (X_{ij}) and update them from their full conditional distributions, treating them exactly the same as the model parameters. Therefore, each sample of the regression coefficient β is updated using a complete set of covariates, but the imputed covariates vary across iterations following their posterior distribution. The missing values are thus effectively treated as nuisance parameters, and the analysis produces both the posterior predictive distribution of the missing values but more importantly the posterior distribution of the regression coefficients marginally over the uncertainty in the missing observations, as desired.

Marathon example: In Section 2.1.7, we modelled the covariance of the covariates ($\mathbf{\Sigma}$) using an inverse Wishart model. This model did not assume any structure among the miles, but the posterior mean of the covariance matrix revealed that subsequent miles are highly correlated. Therefore, here we model the standardized (to have mean zero and variance one) covariates using the first-order autoregressive times series model $X_{i1} \sim \text{Normal}(0, \sigma_1^2)$ and $X_{ij+1}|X_{ij} \sim \text{Normal}(\rho X_{ij}, \sigma_2^2)$ for $j \geq 1$. JAGS code for this model is given in Listing 6.3. All hyperparameters have uninformative priors.

Figure 6.10b plots the observed covariates (dots) and the posterior distribution of the missing covariates (boxplots) for two representative runners. Because of the times series model for the missing covariates, the posterior distributions of the missing X_{ij} are close to the speeds for the adjacent miles for both runners. The posterior distributions of the covariates, β_j, are plotted in Figure 6.10c. Only miles 24 and 25 appear to be useful predictors of the speed in the final mile. The posterior variances in this missing-data analysis are much smaller than the posterior variances in the complete-case analysis with $n = 58$ in Figure 6.10d, illustrating the benefit of the missing data model.

6.5 Exercises

1. Since full conditional distributions are used in many MCMC algorithms, it is tempting to specify the model via its conditional

Listing 6.3
JAGS model statement for the missing data analysis of the marathon data.

```
1    # Likelihood
2    for(i in 1:n){
3      Y[i] ~ dnorm(alpha + inprod(X[i,],beta[]),taue)
4    }
5
6    # Missing-data model
7    for(i in 1:n){
8      X[i,1] ~ dnorm(0,tau1)
9      for(j in 2:p){
10       X[i,j] ~ dnorm(rho*X[i,j-1],tau2)
11     }
12   }
13
14   # Priors
15   alpha ~ dnorm(0,0.01)
16   for(j in 1:p){
17     beta[j] ~ dnorm(0,0.01)
18   }
19   taue ~ dgamma(0.1, 0.1)
20   tau1 ~ dgamma(0.1, 0.1)
21   tau2 ~ dgamma(0.1, 0.1)
22   rho  ~ dnorm(0, 0.01)
```

distributions. For example, consider the conditional distributions:

$$Y|X \sim \text{Normal}(aX, 1) \quad \text{and} \quad X|Y \sim \text{Normal}(bY, 1).$$

(a) Select values of a and b so that these full conditional distributions are incompatible, i.e., there is no valid joint distribution for X and Y that gives these full conditional distributions. Argue, but do not formally prove, your assertion that the full conditional distributions are incompatible.

(b) Explain why building a model that produces a valid DAG will always lead to a valid joint distribution.

2. Draw a DAG for the model in Listing 6.3, derive the full conditional posterior distribution for $X_{2,2}$ (i.e., X[2,2], the speed for runner 2 at mile 2), and explain how this would be used in a Gibbs sampler.

3. In this problem we will conduct a meta analysis, i.e., an analysis that combines the results of several studies. The data are from the rmeta package in R:

```
> library(rmeta)
> data(cochrane)
> cochrane
          name ev.trt n.trt ev.ctrl n.ctrl
1      Auckland     36   532      60    538
2         Block      1    69       5     61
3         Doran      4    81      11     63
4         Gamsu     14   131      20    137
5      Morrison      3    67       7     59
6 Papageorgiou      1    71       7     75
7       Tauesch      8    56      10     71
```

The data are from seven randomized trials that evaluate the effect of corticosteroid therapy on neonatal death. For trial $i \in \{1, ..., 7\}$ denote Y_{i0} as the number of events in the N_{i0} control-group patients and Y_{i1} as the number of events in the N_{i1} treatment-group patients.

(a) Fit the model $Y_{ij}|\theta_j \overset{indep}{\sim} \text{Binomial}(N_{ij}, \theta_j)$ with $\theta_0, \theta_1 \sim$ Uniform$(0, 1)$. Can we conclude that the treatment reduces the event rate?

(b) Fit the model $Y_{ij}|\theta_{ij} \overset{indep}{\sim} \text{Binomial}(N_{ij}, \theta_{ij})$ with $\text{logit}(\theta_{ij}) = \alpha_{ij}$ and $\alpha_i = (\alpha_{i0}, \alpha_{i1})^T \overset{iid}{\sim} \text{Normal}(\mu, \Sigma)$, $\mu \sim$ Normal$(0, 10^2 I_2)$, and $\Sigma \sim \text{InvWishart}(3, I_2)$. Summarize the evidence that the treatment reduces the death rate.

(c) Draw a DAG for these two models.

(d) Discuss the advantages and disadvantages of both models.

(e) Which model is preferred for these data?

4. Download the marathon data of Section 6.4 from the course web-page. Let Y_{ij} be the speed of runner i in mile j. Fit the hierarchical model $Y_{i1} \sim \text{Normal}(\mu_i, \sigma_0^2)$ and

$$Y_{ij}|Y_{ij-1} \sim \text{Normal}(\mu_i + \rho_i(Y_{ij-1} - \mu_i), \sigma_i^2),$$

where $\mu_i \overset{iid}{\sim} \text{Normal}(\theta_1, \theta_2)$, $\rho_i \overset{iid}{\sim} \text{Normal}(\theta_3, \theta_4)$, and $\sigma_i^2 \overset{iid}{\sim}$ InvGamma(θ_5, θ_6).

(a) Draw a DAG for this model and give an interpretation for each parameter in the model.

(b) Select uninformative prior distributions for $\theta_1, ..., \theta_6$.

(c) Fit the model in JAGS using three chains each with 25,000 iterations and thoroughly assess MCMC convergence for the θ_j.

(d) Are the data informative about the θ_j? That is, are the posterior distributions more concentrated than the prior distributions?

(e) In light of (c) and (d), are there any simplifications you might consider and if so, how would you compare the full and simplified models?

5. Download the Gambia data

```
> library(geoR)
> data(gambia)
> ?gambia
```

The data consist of 2,035 children that live in 65 villages. For village $v \in \{1, ..., 65\}$, denote n_v as the number of children in the sample, Y_v as the number of children that tested positive for malaria, and p_v as the true probability of testing positive for malaria. We use the spatial model $\alpha_v = \text{logit}(p_v)$, where $\boldsymbol{\alpha} = (\alpha_1, ..., \alpha_{65})^T$ follows a multivariate normal distribution with mean $\text{E}(\alpha_v) = \mu$, variance $\text{V}(\alpha_v) = \sigma^2$, and correlation $\text{Cor}(\alpha_u, \alpha_v) = \exp(-d_{uv}/\rho)$, where d_{uv} is the distance between villages u and v. For priors assume $\mu \sim \text{Normal}(0, 10^2)$, $\sigma^2 \sim \text{InvGamma}(0.1, 0.1)$, and $\rho \sim \text{Uniform}(0, d^*)$ where d^* is the maximum distance between villages.

(a) Specify the data layer, process layer and prior layer for this hierarchical model.

(b) Fit the model using JAGS and assess convergence.

(c) Summarize the data and results using five maps: the sample size n_v, the sample proportion Y_v/n_v, the posterior means of the p_v, the posterior standard deviations of the p_v, and the posterior probabilities that p_v exceeds 0.5.

7

Statistical properties of Bayesian methods

In this chapter we briefly discuss some of the most important concepts of statistical theory as they relate to Bayesian methods. This book is primarily dedicated to the practical application of Bayesian statistical methods and thus this chapter can be skipped on first read. However, even an applied statistician should be familiar with statistical theory at least at a high level to plan and defend their work.

As an example, consider the problem of estimating a normal mean, θ. An estimator is a function that takes the data as input and returns an estimate of the parameter of interest. An estimator of the parameter θ is denoted $\hat{\theta}(\mathbf{Y})$. A natural estimator of a normal mean is the sample mean, $\hat{\theta}(\mathbf{Y}) = \sum_{i=1}^{n} Y_i/n$. However, there are other estimators including the sample median or the trimmed mean. In Section 2.1.3 we discussed Bayesian estimators such as the posterior mean $\hat{\theta}(\mathbf{Y}) = \mathrm{E}(\theta|\mathbf{Y}) = \sum_{i=1}^{n} Y_i/(n+m)$. How to justify that the Bayesian estimator is a better estimator than the sample mean? In what sense is an estimator optimal? Questions such as these motivate the development of statistical theory.

In addition to investigating the properties of specific estimators and determining the settings for which they are preferred, much of statistical theory is dedicated to developing general procedures for deriving estimators with good properties. General estimation procedures include maximum likelihood estimation, the method of moments, estimating equations, and of course, Bayesian methods. Comparisons of these statistical frameworks are often made using frequentist criteria. As we will see, even if the objective if to generate a procedure with good frequentist properties, Bayesian methods often perform well.

In Section 7.1 we focus on Bayesian methods and optimally summarizing the posterior distribution. In Sections 7.2 and 7.3 we study the frequentist properties of Bayesian methods using mathematical and computational tools, respectively.

7.1 Decision theory

Before studying the frequentist properties of Bayesian estimators, we discuss how to optimally summarize the posterior distribution. An appealing feature of a Bayesian analysis is that it produces the full posterior distribution of each parameter. However, in many cases a point estimate (i.e., a single-number estimator) is desired. Common one-number summaries of the posterior used as point estimators are the posterior mean, median or mode. Rather than arbitrarily choosing one of these posterior summaries as the estimator, Bayesian decision theory provides a formal way to select the optimal Bayesian point estimator.

Defining the optimal point estimator requires defining the measure of estimation accuracy that is to be optimized. If θ is the true value of the parameter and $\hat{\theta}(\mathbf{Y})$ is the estimator, then the loss function is defined as $l(\theta, \hat{\theta}(\mathbf{Y}))$. For example, we might choose squared error loss, $l(\theta, \hat{\theta}(\mathbf{Y})) = [\theta - \hat{\theta}(\mathbf{Y})]^2$, or absolute loss $l(\theta, \hat{\theta}(\mathbf{Y})) = |\theta - \hat{\theta}(\mathbf{Y})|$.

The loss function depends on the true value of the parameter, and therefore cannot be evaluated in a real data analysis. From the Bayesian perspective, given the data, uncertainty about the fixed but unknown parameter θ is quantified by its posterior distribution. In this view, θ, and thus $l(\theta, \hat{\theta}(\mathbf{Y}))$, are random variables. The Bayesian risk is the expected (with respect to the posterior distribution of θ) loss, $R(\hat{\theta}(\mathbf{Y})) = \mathrm{E}[l(\theta, \hat{\theta}(\mathbf{Y}))|\mathbf{Y}]$. The *Bayes rule* is the estimator $\hat{\theta}(\mathbf{Y})$ that minimizes Bayesian risk.

For example, assume squared error loss and define the posterior mean as $\bar{\theta} = \mathrm{E}(\theta|\mathbf{Y})$. The Bayesian risk is

$$
\begin{aligned}
R(\hat{\theta}(\mathbf{Y})) &= \mathrm{E}[l(\theta, \hat{\theta}(\mathbf{Y}))|\mathbf{Y}] \\
&= \mathrm{E}\{[\theta - \hat{\theta}(\mathbf{Y})]^2|\mathbf{Y}\} \\
&= \mathrm{E}\{[(\theta - \bar{\theta}) - (\hat{\theta}(\mathbf{Y}) - \bar{\theta})]^2|\mathbf{Y}\} \\
&= \mathrm{E}\{(\theta - \bar{\theta})^2|\mathbf{Y}\} - 2\mathrm{E}\{(\theta - \bar{\theta})(\hat{\theta}(\mathbf{Y}) - \bar{\theta})|\mathbf{Y}\} + \mathrm{E}\{[\hat{\theta}(\mathbf{Y}) - \bar{\theta}]^2|\mathbf{Y}\} \\
&= \mathrm{V}(\theta|\mathbf{Y}) - 2(\hat{\theta}(\mathbf{Y}) - \bar{\theta})\mathrm{E}(\theta - \bar{\theta}|\mathbf{Y}) + [\hat{\theta}(\mathbf{Y}) - \bar{\theta}]^2 \\
&= \mathrm{V}(\theta|\mathbf{Y}) + [\hat{\theta}(\mathbf{Y}) - \bar{\theta}]^2.
\end{aligned}
$$

In the last equation, $\mathrm{E}(\theta - \bar{\theta}|\mathbf{Y}) = \mathrm{E}(\theta) - \bar{\theta} = 0$ by definition. The estimator $\hat{\theta}(\mathbf{Y})$ does not affect the posterior variance $\mathrm{V}(\theta|\mathbf{Y})$, and so under squared error loss, the posterior mean $\hat{\theta}(\mathbf{Y}) = \bar{\theta}$ minimizes Bayesian risk and is the Bayes rule estimator. This justifies the use of the posterior mean to summarize the posterior in cases where squared error loss is reasonable.

Different loss functions give different Bayes rule estimators. Absolute loss gives the posterior median as the Bayes rule estimator. More generally, if over-estimation and under-estimation are weighted differently, a useful loss function is the asymmetric check-loss function

$$l(\theta, \hat{\theta}(\mathbf{Y})) = \begin{cases} (1-\tau)(\hat{\theta}(\mathbf{Y}) - \theta) & \text{if } \hat{\theta}(\mathbf{Y}) > \theta \\ \tau(\theta - \hat{\theta}(\mathbf{Y})) & \text{if } \hat{\theta}(\mathbf{Y}) < \theta \end{cases} \tag{7.1}$$

for $\tau \in (0,1)$. In this loss function, if τ is close to zero, then over-estimation (top row) is penalized more than under-estimation (bottom row). The Bayes rule for this loss function is the τth posterior quantile. If θ is a discrete random variable, then the MAP estimator $\hat{\theta}(\mathbf{Y}) = \arg\max_\theta f(\theta|\mathbf{Y})$ is the Bayes rule under zero-one loss $l(\theta, \hat{\theta}(\mathbf{Y})) = I[\theta \neq \hat{\theta}(\mathbf{Y})]$.

Decision theory can also formalize Bayesian hypothesis testing. Let $H \in \{0,1\}$ denote the true state, with $H = 0$ if the null hypothesis is true and $H = 1$ if the alternative hypothesis is true. A Bayesian analysis produces posterior probabilities $\text{Prob}(H = h|\mathbf{Y})$ for $h = \{0,1\}$ (see Section 5.2). The decision to be made is which hypothesis to select. Let $d(\mathbf{Y}) = 0$ if we select the null hypothesis and $d(\mathbf{Y}) = 1$ if we select the alternative hypothesis.

To determine the Bayes rule for $d(\mathbf{Y})$ we must specify the loss incurred if we select the wrong hypothesis. If we assign a Type I error the loss λ_1 and a Type II error the loss λ_2, the loss function can be written

$$l(H, d(\mathbf{Y})) = \begin{cases} 0 & \text{if } H = d(\mathbf{Y}) \\ \lambda_1 & \text{if } H = 0 \text{ and } d(\mathbf{Y}) = 1 \\ \lambda_2 & \text{if } H = 1 \text{ and } d(\mathbf{Y}) = 0. \end{cases} \tag{7.2}$$

The Bayesian risk is

$$R(d(\mathbf{Y})) = \begin{cases} \lambda_1 \text{Prob}(H = 0|\mathbf{Y}) & \text{if } d(\mathbf{Y}) = 1 \\ \lambda_2 \text{Prob}(H = 1|\mathbf{Y}) & \text{if } d(\mathbf{Y}) = 0. \end{cases} \tag{7.3}$$

Therefore, the Bayes rule is to reject the null hypothesis and conclude that the alternative is true, i.e., select $d(\mathbf{Y}) = 1$, if the posterior probability of the alternative hypothesis exceeds

$$\text{Prob}(H = 1|\mathbf{Y}) > \frac{\lambda_1}{\lambda_1 + \lambda_2}. \tag{7.4}$$

Note that unlike classical hypothesis testing if we swap the roles of the hypotheses the decision rule remains the same. Often the loss of a Type 1 error is assumed to be larger than a Type II error, e.g., $\lambda_1 = 10\lambda_2$, so that the null hypothesis is rejected only if the posterior probability of the alternative hypothesis is near one.

These decision-theory results hold for prediction as well. For example, if predictions are evaluated using squared-error prediction loss, then the mean of the posterior predictive distribution is the Bayes rule. Also, if the response is binary then the Bayes rule for classification is to predict $Y^{pred} = 1$ if the posterior predictive probability $\text{Prob}(Y^{pred} = 1|\mathbf{Y}) > \lambda_1/(\lambda_1 + \lambda_2)$, where λ_1 is the loss for a false positive and λ_2 is the loss of a false negative. In this case, it might be reasonable to set $\lambda_1 = \lambda_2$ and predict $Y^{pred} = 1$ if $\text{Prob}(Y^{pred} = 1|\mathbf{Y}) > 0.5$.

7.2 Frequentist properties

Comparison and evaluation of statistical procedures proposed for general use are commonly evaluated using frequentist criteria. A frequentist evaluation of an estimator focuses on its sampling distribution, i.e., the distribution of $\hat{\theta}(\mathbf{Y})$ over repeated samples of the data $\mathbf{Y}$. The distribution of the data of course depends on the true value of θ, denoted in this discussion of frequentist properties as θ_0. For example, if $Y_i \overset{iid}{\sim} \text{Normal}(\theta_0, \sigma^2)$, then the sample mean $\hat{\theta}(\mathbf{Y}) = \sum_{i=1}^{n} Y_i/n$ is an estimator of θ and its sampling distribution is $\hat{\theta}(\mathbf{Y}) \sim \text{Normal}(\theta_0, \sigma^2/n)$.

There are many ways to compare the sampling distributions of two estimators. For example, we could study the bias

$$\text{Bias}[\hat{\theta}(\mathbf{Y})] = E[\hat{\theta}(\mathbf{Y})] - \theta_0. \tag{7.5}$$

If the bias is positive, this means that on average the estimator over-estimates the parameter and vice versa. With all else equal, an unbiased estimator with bias to equal zero is preferred. However, the bias only evaluates the center of the sample distribution and ignores its spread. Therefore we also study the variance of the sampling distribution $V[\hat{\theta}(\mathbf{Y})]$. For two estimators that are unbiased we prefer low variance because this means the sampling distribution is concentrated around the true parameter; on the other hand, small variance can be undesirable for biased estimators (Figure 7.1). The most common method of comparison is mean squared error

$$MSE[\hat{\theta}(\mathbf{Y})] = E[\hat{\theta}(\mathbf{Y}) - \theta_0]^2 = \text{Bias}[\hat{\theta}(\mathbf{Y})]^2 + \text{Var}[\hat{\theta}(\mathbf{Y})], \tag{7.6}$$

which combines for bias and variance into a single measure.

Evaluating estimators using bias, variance and MSE is complicated by the fact that these summaries depend on the true parameter value, θ_0, and the sample size, n. For example, it may be that an estimator performs well if θ_0 is close to zero and the sample is small, but performs poorly in other settings. Therefore it is important to evaluate procedure over a range of settings. To deal with the sample size, we often consider asymptotic arguments with $n \to \infty$. A method is asymptotically unbiased if its bias converges to zero as the sample size goes to infinity. An estimator is said to be consistent if it converges in

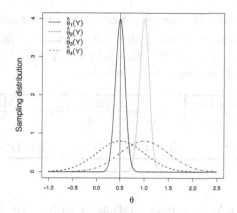

FIGURE 7.1
Hypothetical sampling distributions. Of the four hypothetical sampling distributions, the first two are unbiased and the last two are biased, and the first and third have small variance while the second and fourth have large variance. The true value θ_0 is denoted with the vertical line at $\theta = 0.5$. The mean squared errors of the four estimators are 0.01, 0.25, 0.26 and 0.50, respectively.

probability to the true value, i.e., the probability of the estimator being within ϵ (for any ϵ) of the true value increases to one as the sample size increases to infinity. An asymptotically unbiased estimator is consistent if its variance decreases to zero with the sample size.

Frequentist evaluation of statistical methods extends to interval estimation and testing. For Bayesian credible intervals we evaluate their frequentist coverage probability, i.e., the probability that they include the true value when applied repeatedly to many datasets. Similarly, for a Bayesian testing procedure we compute the probability of making Type I and Type II errors when the test is applied to many random datasets.

7.2.1 Bias-variance tradeoff

For difficult problems such as analyses with many parameters or regression with correlated predictors, adding prior information can stabilize the statistical analysis by reducing the variance of the estimator. On the other hand, if the prior information is erroneous then this can lead to bias. The balance between these two consequences of incorporating prior information can be formalized by studying the bias-variance tradeoff. Recall that the mean squared error is the sum of the variance and the squared bias,

$$\text{MSE}[\hat{\theta}(\mathbf{Y})] = \text{Bias}[\hat{\theta}(\mathbf{Y})]^2 + \text{Var}[\hat{\theta}(\mathbf{Y})]. \tag{7.7}$$

TABLE 7.1
Bias-variance tradeoff for a normal-mean analysis. Assuming $Y_i, ..., Y_n \overset{iid}{\sim}$ Normal(θ, σ^2), this table gives the bias, variance and mean squared error (MSE) of the estimators $\hat{\theta}_1(\mathbf{Y}) = \bar{Y}$ and $\hat{\theta}_2(\mathbf{Y}) = c\bar{Y}$, where $c = n/(n+m) \in [0,1]$ and θ_0 is the true value.

Estimator	Bias	Variance	MSE
$\hat{\theta}_1(\mathbf{Y})$	0	$\frac{\sigma^2}{n}$	$\frac{\sigma^2}{n}$
$\hat{\theta}_2(\mathbf{Y})$	$(c-1)\theta_0$	$\frac{c^2\sigma^2}{n}$	$(c-1)^2\theta_0^2 + \frac{c^2\sigma^2}{n}$

Therefore it is possible for a biased estimator to have smaller MSE than an unbiased estimator if the reduction in variance is large enough to offset the squared bias.

The optimal estimator as given by the bias-variance tradeoff often depends on the true value of the parameter and the sample size. As a concrete example, consider the simple case with $Y_1, ..., Y_n \overset{iid}{\sim}$ Normal(θ, σ^2). For mathematical simplicity we assume σ is fixed. We compare two estimators:

(1) $\hat{\theta}_1(\mathbf{Y}) = \frac{1}{n}\sum_{i=1}^{n} Y_i = \bar{Y}$

(2) $\hat{\theta}_2(\mathbf{Y}) = \frac{1}{n+m}\sum_{i=1}^{n} Y_i = c\bar{Y}$

where $c = n/(n+m) \in [0,1]$. The first estimator is the usual sample mean (i.e., the MLE or posterior mean under Jeffreys prior) and the second is the posterior mean assuming prior $\theta \sim$ Normal$(0, \sigma^2/m)$.

Table 7.1 gives the bias, variance and MSE of the two estimators. The sample mean is unbiased but always has larger variance than the Bayesian estimator. The relative MSE is

$$RMSE = \frac{MSE[\hat{\theta}_2(\mathbf{Y})]}{MSE[\hat{\theta}_1(\mathbf{Y})]} = \frac{nm^2(\theta_0/\sigma)^2 + n^2}{(n+m)^2}. \tag{7.8}$$

The Bayesian procedure is preferred when this ratio is less than one. When $\theta_0 = 0$, the prior mean, then the ratio is less than one for all n and m. That is, when the prior mean is the true value, the Bayesian estimator is preferred over the sample mean. The Bayesian estimator can be preferred even if the prior mean is not exactly the true value. As long as the true value is close to zero,

$$|\theta_0| < \sigma\sqrt{\frac{1}{n} + \frac{2}{m}} = B(n, m) \tag{7.9}$$

the Bayesian estimator is preferred. The bound $B(n,m)$ shrinks to zero as the sample size increases and so in this case the advantage of the Bayesian approach is for small datasets. As $n \to \infty$ or $m \to 0$ the RMSE converges to one, and thus for large sample sizes or uninformative priors, the two estimators perform similarly.

7.2.2 Asymptotics

In Section 3.1.3 we discussed the Bayesian central limit theorem that states that under general conditions the posterior converges to a normal distribution as the sample size increases. It can also be shown that under these conditions and assuming the prior includes the true value that the sampling distribution of the posterior mean $\hat{\boldsymbol{\theta}}_B = \mathrm{E}(\boldsymbol{\theta}|\mathbf{Y})$ converges (as $n \to \infty$) to the sampling distribution of the maximum likelihood estimator

$$\hat{\boldsymbol{\theta}}_B \sim \mathrm{Normal}\left(\boldsymbol{\theta}_0, \hat{\boldsymbol{\Sigma}}_{MLE}\right) \tag{7.10}$$

where $\boldsymbol{\theta}_0$ is the true value and $\hat{\boldsymbol{\Sigma}}_{MLE}$ is defined in Section 3.1.3. Therefore, the posterior mean is asymptotically unbiased. Further, it follows from classic results for maximum likelihood estimators that its variance decreases to zero as the sample size increases and thus the posterior mean is a consistent estimator for $\boldsymbol{\theta}$ in essentially the same conditions as the maximum likelihood estimator.

An asymptotic property that is uniquely Bayesian is posterior consistency. For a given dataset $\mathbf{Y}$, the posterior probability that θ is within ϵ of the true value is

$$\mathrm{Prob}(||\theta - \theta_0|| < \epsilon|\mathbf{Y}). \tag{7.11}$$

A Bayesian procedure is said to possess posterior consistency if this probability is assured to converge to one as the sample size increases (Figure 7.2). Appendix A.3 provides a proof that the posterior distribution is consistent for parameters with discrete support under very general conditions on the likelihood and prior, and posterior consistency has been established for most finite-dimensional problems and many Bayesian nonparametric methods. For a more thorough discussion of posterior consistency see [37].

Both the Bayesian CLT and posterior consistency result in Appendix A.3 hold for any prior as long as the prior does not change with the sample size and has positive mass/density around the true value. *This confirms the argument that for large datasets, any reasonable prior distribution should lead to the same statistical inference and that this posterior will converge to the true value.*

7.3 Simulation studies

For simple models such as the normal mean example in Section 7.2.1 it is possible to derive the frequentist properties of an estimator using algebra. However, for more complicated models, a purely mathematical study is impossible. Especially in these complicated settings, it is important to understand

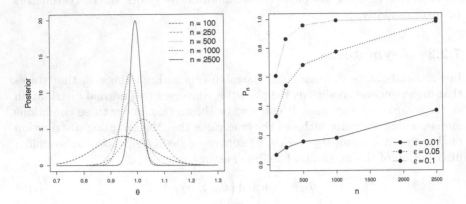

FIGURE 7.2
Illustration of posterior consistency. The data are generated as
$Y_i, ..., Y_N \overset{iid}{\sim} \text{Normal}(\theta_0, 1)$ with $N = 2,500$ and $\theta_0 = 1$, and we fit the
Bayesian model with flat (improper) prior for the mean and variance fixed
at one using the first n observations $\mathbf{Y}_n = (Y_1, ..., Y_n)$. The left panel plots
the posterior $f(\theta|\mathbf{Y}_n)$ by n, and the right panel plots the corresponding pos-
terior probability $P_n = \text{Prob}(|\theta - \theta_0| < \epsilon|\mathbf{Y}_n)$.

the operating characteristics of a statistical procedure, and a simulation study
is a general way to carry out this evaluation.

Just as MCMC is used to approximate complicated posterior distributions, simulation studies use Monte Carlo sampling to approximate complicated sampling distributions. To approximate, say, the mean squared error of
an estimator we generate S independent and identically distributed datasets
given the true parameter value θ_0 and sample size n, and then compute the
estimator for each simulated dataset. If we denote the S simulated datasets as
$\mathbf{Y}_1, ..., \mathbf{Y}_S$, then $\hat{\theta}(\mathbf{Y}_1), ..., \hat{\theta}(\mathbf{Y}_S)$ are S draws from the sampling distribution
of the estimator $\hat{\theta}(\mathbf{Y})$. The mean squared error is then approximated as

$$\text{MSE}[\hat{\theta}(\mathbf{Y})] \approx \frac{1}{S} \sum_{s=1}^{S} [\hat{\theta}(\mathbf{Y}_s) - \theta_0]^2. \quad (7.12)$$

Other summaries of the sampling distribution such as bias and coverage are
computed similarly. Of course, this is only a Monte Carlo estimate of the
true MSE and a different simulation experiment will give a different estimate.
Therefore, the approximation should be accompanied by a standard error,
$s_{MSE}/\sqrt{S}$, where s_{MSE} is the sample standard deviation of the S squared
errors, $[\hat{\theta}(\mathbf{Y}_1) - \theta_0]^2, ..., [\hat{\theta}(\mathbf{Y}_S) - \theta_0]^2$.

A typical simulation study will generate data from a few values of θ_0 and

n to understand when the method performs well and when it does not. Note that if $\hat{\theta}(\mathbf{Y})$ is a Bayesian estimator, say the posterior mean, then computing each $\hat{\theta}(\mathbf{Y}_s)$ may require MCMC and thus Bayesian simulation experiments may need many applications of MCMC and can be time consuming (running the S chains in parallel is obviously helpful). For a thorough description of simulation studies, see [13] (Chapter 9).

As an example, we conduct a simulation study to compare ordinary least squares (OLS) regression with Bayesian LASSO regression (BLR; Section 4.2). The sampling distribution of the OLS estimator is known (multivariate student-t) but the sampling distribution of the posterior mean under the BLR model is quite complicated and difficult to study without simulation. We generate $X_{ij} \overset{iid}{\sim} \text{Normal}(0,1)$ and $Y_i|\mathbf{X}_i \sim \text{Normal}(\sum_{j=1}^{p} X_{ij}\beta_{j0}, \sigma_0^2)$. The data are generated with true values $\sigma_0 = 1$ and the first p_0 elements of $\boldsymbol{\beta}_0$ equal to zero and the final p_1 elements equal to one, so that $p = p_0 + p_1 = 20$. We generate $S = 100$ datasets each from six combinations of n, p_0 and p_1 using the R code in Listing 7.1. Each dataset is analyzed using least squares (the lm function in R) and Bayesian LASSO (the BLR function in R with default values). For dataset $s = 1, ..., S$, let $\hat{\beta}_{js}$ be the estimate of β_j (either the least squares solution for OLS or the posterior mean for BLR) and v_{js} its estimated variance (either the squared standard error for OLS or the posterior variance for BLR). Methods are compared using

$$\text{Bias} \;=\; \frac{1}{Sp}\sum_{s=1}^{S}\sum_{j=1}^{p}(\hat{\beta}_{js} - \beta_{j0})$$

$$\text{Variance} \;=\; \frac{1}{Sp}\sum_{s=1}^{S}\sum_{j=1}^{p}v_{js}$$

$$\text{MSE} \;=\; \frac{1}{Sp}\sum_{s=1}^{S}\sum_{j=1}^{p}(\hat{\beta}_{js} - \beta_{j0})^2$$

$$\text{Coverage} \;=\; \frac{1}{Sp}\sum_{s=1}^{S}\sum_{j=1}^{p}I(|\hat{\beta}_{js} - \beta_{j0}| < 2\sqrt{v_{js}}).$$

For coverage, we approximate the posterior distribution as Gaussian to avoid saving all posterior samples. Results are averaged across covariates and standard errors are omitted to make the presentation concise. It may also be interesting to study these coefficients separately, in particular, to evaluate performance separately for null and active covariates.

Table 7.2 shows the results. For the small sample size ($n = 40$), the Bayesian method gives a large reduction in variance and thus mean squared error in the first two cases with more null than active covariates. In these cases the prior information that many of the coefficients are near zero is valid and improves the stability of the algorithm. However, when this prior information is wrong in case three and all covariates are active the BLR

Listing 7.1
R simulation study code.

```r
1   # Set up the simulation
2   library(BLR)
3   n       <- 25       # Sample size
4   p_null <- 15        # Number of null covariates
5   p_act  <- 5         # Number of active covariates
6   nsims  <- 100       # Number of simulated datasets
7   sigma  <- 1         # True value of sigma
8   beta   <- c(rep(0,p_null),rep(1,p_act)) # True beta
9
10  # Define matrices to store the results
11  p     <- p_null + p_act
12  EST1 <- VAR1 <- matrix(0,nsims,p)
13  EST2 <- VAR2 <- matrix(0,nsims,p)
14
15  # Start the simulation
16  for(sim in 1:nsims){
17   set.seed(sim*1234)
18    # Generate a dataset
19     X <- matrix(rnorm(n*p),n,p)
20     Y <- X%*%beta+rnorm(n,0,sigma)
21
22    # Fit ordinary least squares
23      ols         <- summary(lm(Y~X))$coef[-1,]
24     EST1[sim,] <- ols[,1]
25     VAR1[sim,] <- ols[,2]^2
26
27    # Fit the Bayesian LASSO
28      blr         <- BLR(y=Y,XL=X)
29     EST2[sim,] <- blr$bL
30     VAR2[sim,] <- blr$SD.bL^2
31   }
32
33  # Compute the results
34  E    <- sweep(EST1,2,beta,"-")
35  MSE  <- mean(E^2)
36  BIAS <- mean(E)
37  VAR  <- mean(VAR1)
38  COV  <- mean(abs(E/sqrt(VAR1))<2)
39
40  E    <- sweep(EST2,2,beta,"-")
41  MSE  <- c(MSE,mean(E^2))
42  BIAS <- c(BIAS,mean(E))
43  VAR  <- c(VAR,mean(VAR2))
44  COV  <- c(COV,mean(abs(E/sqrt(VAR2))<2))
45
46  out <- cbind(BIAS,VAR,MSE,COV)
```

TABLE 7.2
Simulation study results. The simulation study compares ordinary least squares ("OLS") with Bayesian LASSO regression ("BLR") for estimating regression coefficients in terms of bias, variance, mean squared error ("MSE") and coverage of 95% intervals (all metrics are averaged over covariates and datasets). The simulations vary based on the sample size (n), the number of null covariates (p_0) and the number of active covariates (p_1). All values are multiplied by 100.

n	p_0	p_1	Bias OLS	Bias BLR	Variance OLS	Variance BLR	MSE OLS	MSE BLR	Coverage OLS	Coverage BLR
40	20	0	-1.65	-0.08	5.59	0.19	5.40	0.03	94.7	100.0
	15	5	0.63	-3.14	5.38	3.47	5.71	3.45	93.8	96.0
	0	20	-0.88	-11.71	5.59	7.09	5.40	9.47	93.7	91.6
100	20	0	-0.43	-0.04	1.28	0.09	1.17	0.02	95.8	100.0
	15	5	0.44	-0.56	1.22	1.02	1.27	0.98	94.5	95.5
	0	20	0.11	-1.27	1.33	1.36	1.22	1.26	96.0	95.6

method has larger bias and thus MSE than OLS and the empirical coverage of the Bayesian credible sets dips below the nominal level. For the large samples size cases ($n = 100$), these same trends are apparent but the differences between methods are smaller, as expected.

7.4 Exercises

1. Assume $Y|\mu \sim \text{Normal}(\mu, 2)$ and $\mu \sim \text{Normal}(0, 2)$ (i.e., $n = 1$). The objective is to test the null hypothesis $H_0 : \mu \leq 0$ versus the alternative hypothesis that $H_1 : \mu > 0$. We will reject H_0 if $\text{Prob}(H_1|Y) > c$.

   (a) Compute the posterior of μ.
   (b) What is the optimal value of c if Type I and Type II errors have the same costs?
   (c) What is the optimal value of c if a Type I error costs ten times more than a Type II error?
   (d) Compute the Type I error rate of the test as a function of c.
   (e) How would you pick c to control Type I error at 0.05?

2. Given data from a small pilot study, your current posterior probability that the new drug your company has developed is more effective

than the current treatment is $\theta \in [0, 1]$. Your company is considering to run a large clinical trial to confirm that your drug is indeed preferred. If you run the trial it will cost \$X. If in fact your drug is better, then the probability that you will confirm this in the trial is 80%; if in fact your drug is not better there is still a 5% chance the trial will conclude it is better. If the trial suggests your drug is preferred, you will make \$cX. For which values of θ and c would you initiate the trial?

3. Assume $Y|\theta \sim \text{Binomial}(n, \theta)$. Consider two estimators of θ: the sample proportion $\hat{\theta}_1(Y) = Y/n$ and the posterior mean under the Jeffreys' prior $\hat{\theta}_2(Y) = (Y + 1/2)/(n + 1/2)$.

   (a) Compute the bias, variance and MSE for each estimator and comment on the bias-variance tradeoff.

   (b) In terms of n and the true value θ_0, when is $\hat{\theta}_2(Y)$ preferred?

4. Assume $Y_1, ..., Y_n|\sigma^2 \overset{iid}{\sim} \text{Normal}(0, \sigma^2)$ and $\sigma^2 \sim \text{InvGamma}(a, b)$.

   (a) Give an expression for two estimators of σ^2: the posterior mean and the posterior mode.

   (b) Plot the posterior density functions and the give the two estimators for a sample with $n = 10$ and $\sum_{i=1}^{n} Y_i^2 = 200$ and $a = b = 0.001$.

   (c) Argue that both estimators are consistent (i.e., their MSE goes to zero as n increases) for any a and b.

5. Assume $Y|\theta \sim \text{Binomial}(n, \theta)$ and $\theta \sim \text{Beta}(1/2, 1/2)$. Use a simulation study to compute the empirical coverage of the equal-tailed 95% credible set for $n \in \{1, 5, 10, 25\}$ and true value $\theta_0 \in \{0.05, 0.10, ..., 0.50\}$. Comment on the frequentist properties of the Bayesian credible set.

6. A paper reports the results of a logistic regression analysis using maximum likelihood estimation. They estimate β_1 to be $\hat{\beta}_1 = 2.1$, with the 95% confidence interval $(0.42, 3.17)$. Further, the p-value for the test that $H_0 : \beta_1 = 0$ versus $H_1 : \beta_1 \neq 0$ is 0.01. Conduct a Bayesian analysis using only this information and asymptotic arguments.

7. You are designing a study to estimate the success probability of a new marketing strategy. When the data have been collected, you will analyze them using the model $Y|\theta \sim \text{Binomial}(n, \theta)$ and $\theta \sim \text{Uniform}(0, 1)$. Before collecting any data, you suspect (based on past studies) that $\theta \approx 0.4$. Which value of n should be used to ensure the posterior standard deviation will be approximately 0.01?

8. Suppose $Y_1, ..., Y_n \overset{iid}{\sim} f(y|\eta)$ where f is the canonical exponential family

$$f(y|\eta) \propto \exp\{y\eta - \psi(\eta)\}$$

for some known function ψ. We are interested in estimating the parameter $\theta = E(Y_i|\eta) = d\psi(\eta)/d\eta$.

(a) Show that the Gaussian density with variance fixed at one and the Poisson density can be written in this form.

(b) Obtain the maximum likelihood estimator of θ. Is it unbiased for θ?

(c) Find a class of conjugate priors for η.

(d) Obtain the Bayes estimator of μ using a conjugate prior and squared error loss.

(e) Is there any (proper) conjugate prior for which the Bayes estimator that you obtained in (d) is unbiased? Justify your answer.

Appendices

A.1: Probability distributions

Univariate discrete

In the following plots the probability mass functions for several combinations of parameters are denoted with points; lines connect the points for visualization, but the probability is non-zero only at the points.

Discrete uniform

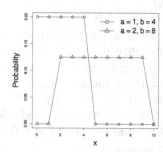

Notation: $X \sim \text{DiscreteUniform}(a, b)$
Support: $X \in \{a, a+1, ..., b\}$
Parameters: $a, b \in \{..., -1, 0, 1, ...\}$ with $a < b$
PMF: $1/(b - a + 1)$
Mean: $(a + b)/2$
Variance: $[(b - a + 1)^2 - 1]/12$
Notes: The discrete uniform can be applied to any finite set. For example, we could say that X is distributed uniformly over the set $\{1/10, 2/10, ..., 10/10\}$.

Binomial

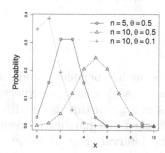

Notation: $X \sim \text{Binomial}(n, \theta)$
Support: $X \in \{0, 1, ..., n\}$
Parameters: $n \in \{1, 2, ...\}$, $\theta \in [0, 1]$
PMF: $\binom{n}{x}\theta^x(1 - \theta)^{n-x}$
Mean: $n\theta$
Variance: $n\theta(1 - \theta)$
Notes: If X is the number of successes in n independent trials each with success probability θ, then $X \sim \text{Binomial}(n, \theta)$; if $n = 1$ then $X \sim \text{Bernoulli}(\theta)$.

Beta-binomial

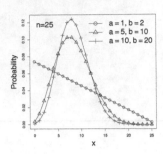

Notation: $X \sim \text{BetaBinomial}(n, a, b)$
Support: $X \in \{0, 1, ..., n\}$
Parameters: $n \in \{1, 2, ...\}$, $a, b > 0$
PMF: $\frac{\Gamma(n+1)\Gamma(x+a)\Gamma(n-x+b)\Gamma(a+b)}{\Gamma(x+1)\Gamma(n-x+1)\Gamma(n+a+b)\Gamma(a)\Gamma(b)}$
Mean: $na/(a+b)$
Variance: $nab(a+b+n)/[(a+b)^2(a+b+1)]$
Notes: If $X|\theta \sim \text{Binomial}(n, \theta)$ and $\theta \sim \text{Beta}(a, b)$, then $X \sim \text{BetaBinomial}(n, a, b)$. If $a = b = 1$ then $X \sim \text{DiscreteUniform}(0, n)$.

Negative Binomial

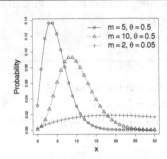

Notation: $X \sim \text{NegBinomial}(\theta, m)$
Support: $X \in \{0, 1, 2, ...\}$
Parameters: $m > 0$, $\theta \in [0, 1]$
PMF: $\binom{x+m-1}{x}\theta^m(1-\theta)^x$
Mean: $m(1-\theta)/\theta$
Variance: $m(1-\theta)/\theta^2$
Notes: In a sequence of independent trials each with success probability θ, if X is the number of failures that occur before the mth success (assuming m is an integer), then $X \sim \text{NegBinomial}(\theta, m)$; if $m = 1$ then $X \sim \text{Geometric}(\theta)$. (The distribution can also be defined with m as the number of failures, but we use the JAGS parameterization.)

Poisson

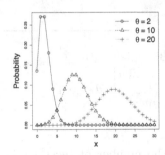

Notation: $X \sim \text{Poisson}(\theta)$
Support: $X \in \{0, 1, 2, ...\}$
Parameters: $\theta > 0$
PMF: $\frac{\theta^x \exp(-\theta)}{x!}$
Mean: θ
Variance: θ
Notes: If events occur independently and uniformly over time (space) with the expected number of events in a given time interval (region) equal to θ, then the number of events that occur in the interval (region) follows a $\text{Poisson}(\theta)$ distribution.

Multivariate discrete

Multinomial

Notation: $\mathbf{X} = (X_1, ..., X_p) \sim \text{Multinomial}(n, \boldsymbol{\theta})$

Support: $X_j \in \{0, 1, ..., n\}$ with $\sum_{j=1}^{p} X_j = n$

Parameters: $\boldsymbol{\theta} = (\theta_1, ..., \theta_p)$ with $\theta_j \in [0, 1]$ and $\sum_{j=1}^{p} \theta_j = 1$

PMF: $\frac{n!}{\prod_{j=1}^{p} x_j!} \prod_{j=1}^{p} \theta_j^{x_j}$

Mean: $\text{E}(X_j) = n\theta_j$

Variance: $\text{V}(X_j) = n\theta_j(1 - \theta_j)$

Covariance: $\text{Cov}(X_j, X_k) = -n\theta_j\theta_k$

Marginal distributions: $X_j \sim \text{Binomial}(n, \theta_j)$

Notes: If n independent trials each have p possible outcomes with the probability of outcome j being θ_j and X_j is the number of the trials that result in outcome j, then $\mathbf{X} = (X_1, ..., X_p) \sim \text{Multinomial}(n, \boldsymbol{\theta})$.

Univariate continuous

Uniform

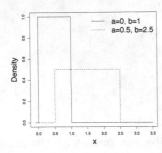

Notation: $X \sim \text{Uniform}(a, b)$

Support: $X \in [a, b]$

Parameters: $-\infty < a < b < \infty$

PDF: $\frac{1}{b-a}$

Mean: $(a + b)/2$

Variance: $(b - a)^2/12$

Notes: If $X_1, X_2 \overset{iid}{\sim} \text{Uniform}(0, 1)$ then $\sqrt{-2 \log(X_1)} \cos(2\pi X_2) \sim \text{Normal}(0, 1)$; if $X \sim \text{Uniform}(0, 1)$ and F is a continuous CDF, then $F^{-1}(X)$ has CDF F.

Beta

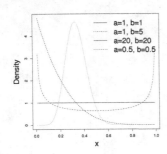

Notation: $X \sim \text{Beta}(a, b)$

Support: $X \in [0, 1]$

Parameters: $a > 0$, $b > 0$

PDF: $\frac{\Gamma(a+b)}{\Gamma(a)\Gamma(b)} x^{a-1}(1 - x)^{b-1}$

Mean: $\frac{a}{a+b}$

Variance: $\frac{ab}{(a+b)^2(a+b+1)}$

Notes: If $a = b = 1$ then $X \sim \text{Uniform}(0, 1)$; $1 - X \sim \text{Beta}(b, a)$.

Gamma

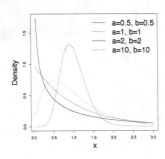

Notation: $X \sim \text{Gamma}(a, b)$

Support: $X \in [0, \infty]$

Parameters: shape $a > 0$, scale $b > 0$

PDF: $\frac{b^a}{\Gamma(a)} x^{a-1} \exp(-bx)$

Mean: a/b

Variance: a/b^2

Notes: $cX \sim \text{Gamma}(a, b/c)$; if $a = 1$ then $X \sim \text{Exponential}(b)$; if $a = \nu/2$ and $b = 1/2$ then $X \sim \text{Chi-squared}(\nu)$; $1/X \sim \text{InvGamma}(a, b)$.

Inverse gamma

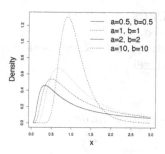

Notation: $X \sim \text{InvGamma}(a, b)$
Support: $X \in [0, \infty]$
Parameters: shape $a > 0$, scale $b > 0$
PDF: $\frac{b^a}{\Gamma(a)} x^{-a-1} \exp(-b/x)$
Mean: $\frac{b}{a-1}$ (if $a > 1$)
Variance: $\frac{b^2}{(a-1)^2(a-2)}$ (if $a > 2$)
Notes: $cX \sim \text{InvGamma}(a, cb)$; $1/X \sim \text{Gamma}(a, b)$.

Normal/Gaussian

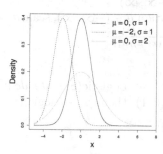

Notation: $X \sim \text{Normal}(\mu, \sigma^2)$
Support: $X \in (-\infty, \infty)$
Parameters: location $\mu \in (-\infty, \infty)$, scale $\sigma > 0$
PDF: $\frac{1}{\sqrt{2\pi}\sigma} \exp\left[-\frac{(x-\mu)^2}{2\sigma^2}\right]$
Mean: μ
Variance: σ^2
Notes: $c + dX \sim \text{Normal}(c + d\mu, d^2\sigma^2)$; if $\mu = 0$ and $\sigma^2 = 1$ then X follows the standard normal distribution.

Student's t

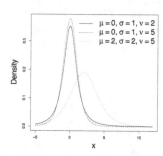

Notation: $X \sim t_\nu(\mu, \sigma^2)$
Support: $X \in (-\infty, \infty)$
Parameters: location $\mu \in (-\infty, \infty)$, scale $\sigma > 0$, degrees of freedom $\nu > 0$
PDF: $\frac{\Gamma(\frac{\nu+1}{2})}{\Gamma(\nu/2)\sqrt{\nu\pi}\sigma}\left[1 + \frac{(x-\mu)^2}{\nu\sigma^2}\right]^{-(\nu+1)/2}$
Mean: μ (if $\nu > 1$)
Variance: $\sigma^2 \frac{\nu}{\nu-2}$ (if $\nu > 2$)
Notes: $c + dX \sim t_\nu(c + d\mu, d^2\sigma^2)$; if $\mu = 0$ and $\sigma^2 = 1$ then X follows the standard t distribution; if $Z \sim \text{Normal}(0, 1)$ independent of $W \sim \text{Gamma}(\nu/2, 1/2)$ then $\mu + \sigma Z/\sqrt{W/\nu} \sim t_\nu(\mu, \sigma^2)$; if $\nu = 1$ then X follows the Cauchy distribution; X is approximately $\text{Normal}(\mu, \sigma^2)$ for large ν.

Laplace/Double exponential

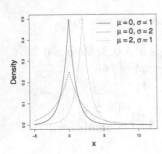

Notation: $X \sim \mathrm{DE}(\mu, \sigma)$
Support: $X \in (-\infty, \infty)$
Parameters: location $\mu \in (-\infty, \infty)$, scale $\sigma > 0$
PDF: $\frac{1}{2\sigma} \exp\left(-\frac{|x-\mu|}{\sigma}\right)$
Mean: μ
Variance: $2\sigma^2$
Notes: $c + dX \sim \mathrm{DE}(c + d\mu, d\sigma)$.

Logistic

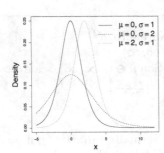

Notation: $X \sim \mathrm{Logistic}(\mu, \sigma)$
Support: $X \in (-\infty, \infty)$
Parameters: location $\mu \in (-\infty, \infty)$, scale $\sigma > 0$
PDF: $\frac{1}{\sigma} \frac{\exp[-(x-\mu)/\sigma]}{\{1+\exp[-(x-\mu)/\sigma]\}^2}$
Mean: μ
Variance: $\pi^2 \sigma^2 / 3$
Notes: $c + dX \sim \mathrm{Logistic}(c + d\mu, d\sigma)$; if $U \sim \mathrm{Uniform}(0,1)$ then $\mu + \sigma \mathrm{logit}(U) \sim \mathrm{Logistic}(\mu, \sigma)$.

Multivariate continuous

Multivariate normal

Notation: $\mathbf{X} = (X_1, ..., X_p)^T \sim \text{Normal}(\boldsymbol{\mu}, \boldsymbol{\Sigma})$
Support: $X_j \in (-\infty, \infty)$
Parameters: mean vector $\boldsymbol{\mu} = (\mu_1, ..., \mu_p)$ with $\mu_j \in (-\infty, \infty)$ and $p \times p$ positive definite covariance matrix $\boldsymbol{\Sigma}$
PDF: $(2\pi)^{-p/2} |\boldsymbol{\Sigma}|^{-1/2} \exp[-\frac{1}{2}(\mathbf{X} - \boldsymbol{\mu})^T \boldsymbol{\Sigma}^{-1}(\mathbf{X} - \boldsymbol{\mu})]$
Mean: $\text{E}(X_j) = \mu_j$
Variance: $\text{V}(X_j) = \sigma_j^2$ where σ_j^2 is the (j, j) element of $\boldsymbol{\Sigma}$
Covariance: $\text{Cov}(X_j, X_k) = \sigma_{jk}$ where σ_{jk} is the (j, k) element of $\boldsymbol{\Sigma}$
Marginal distributions: $X_j \sim \text{Normal}(\mu_j, \sigma_j^2)$
Notes: For q-vector $\mathbf{a}$ and $q \times p$ matrix $\mathbf{b}$, $\mathbf{a} + \mathbf{bX} \sim \text{Normal}(\mathbf{a} + \mathbf{b}\boldsymbol{\mu}, \mathbf{b}\boldsymbol{\Sigma}\mathbf{b}^T)$.

Multivariate t

Notation: $\mathbf{X} = (X_1, ..., X_p)^T \sim t_\nu(\boldsymbol{\mu}, \boldsymbol{\Sigma})$
Support: $X_j \in (-\infty, \infty)$
Parameters: location $\boldsymbol{\mu} = (\mu_1, ..., \mu_p)$ with $\mu_j \in (-\infty, \infty)$, $p \times p$ positive definite matrix $\boldsymbol{\Sigma}$ and degrees of freedom $\nu > 0$
PDF: $\frac{\Gamma(\nu/2+p/2)}{\Gamma(\nu/2)(\nu\pi)^{p/2}} |\boldsymbol{\Sigma}|^{-1/2} \left[1 + \frac{1}{\nu}(\mathbf{X} - \boldsymbol{\mu})^T \boldsymbol{\Sigma}^{-1}(\mathbf{X} - \boldsymbol{\mu})\right]^{-(\nu+p)/2}$
Mean: $\text{E}(X_j) = \mu_j$ (if $\nu > 1$)
Variance: $\text{V}(X_j) = \frac{\nu}{\nu-2}\sigma_j^2$ where σ_j^2 is the (j, j) element of $\boldsymbol{\Sigma}$ (if $\nu > 2$)
Covariance: $\text{Cov}(X_j, X_k) = \frac{\nu}{\nu-2}\sigma_{jk}$ where σ_{jk} is the (j, k) element of $\boldsymbol{\Sigma}$ (if $\nu > 2$)
Marginal distributions: $X_j \sim t_\nu(\mu_j, \sigma_j^2)$
Notes: For q-vector $\mathbf{a}$ and $q \times p$ matrix $\mathbf{b}$, $\mathbf{a} + \mathbf{bX} \sim t_\nu(\mathbf{a} + \mathbf{b}\boldsymbol{\mu}, \mathbf{b}\boldsymbol{\Sigma}\mathbf{b}^T)$; $\mathbf{X}$ is approximately $\text{Normal}(\boldsymbol{\mu}, \boldsymbol{\Sigma})$ for large ν; if $\mathbf{X}|W \sim \text{Normal}(\mathbf{0}, \boldsymbol{\Sigma}/W)$ and $W \sim \text{Gamma}(\nu/2, 1/2)$, then $\mathbf{X} \sim t_\nu(\boldsymbol{\mu}, \boldsymbol{\Sigma})$.

Dirichlet

Notation: $\mathbf{X} = (X_1, ..., X_p) \sim \text{Dirichlet}(\boldsymbol{\theta})$
Support: $X_j \in [0, 1]$ with $\sum_{j=1}^p X_j = 1$
Parameters: $\boldsymbol{\theta} = (\theta_1, ..., \theta_p)$ with $\theta_j > 0$
PDF: $\frac{\Gamma(\sum_{j=1}^p \theta_j)}{\prod_{j=1}^p \Gamma(\theta_j)} \prod_{j=1}^p x_j^{\theta_j-1}$
Mean: $\text{E}(X_j) = \theta_j/(\sum_{k=1}^p \theta_k)$
Variance: $\text{V}(X_j) = \frac{\theta_j(\sum_{k \neq j} \theta_k)}{(\sum_{k=1}^p \theta_k)^2(1+\sum_{k=1}^p \theta_k)}$
Covariance: $\text{Cov}(X_j, X_k) = \frac{-\theta_j\theta_k}{(\sum_{k=1}^p \theta_k)^2(1+\sum_{k=1}^p \theta_k)}$
Marginal distributions: $X_j \sim \text{Beta}(\theta_j, \sum_{k \neq j} \theta_k)$
Notes: If $W_j \overset{indep}{\sim} \text{Gamma}(\theta_j, b)$ and $X_j = W_j/(\sum_{k=1}^p W_k)$ then $\mathbf{X} = (X_1, ..., X_p) \sim \text{Dirichlet}(\boldsymbol{\theta})$.

Wishart

Notation: $\mathbf{X} \sim \text{Wishart}(\nu, \boldsymbol{\Omega})$

Support: $\mathbf{X} = \{X_{jk}\}$ is a $p \times p$ symmetric positive definite matrix

Parameters: degrees of freedom $\nu > p-1$ and $p \times p$ symmetric positive definite matrix $\boldsymbol{\Omega} = \{\Omega_{jk}\}$

PDF: $\frac{1}{2^{p\nu}|\boldsymbol{\Omega}|^{\nu/2}\Gamma_p(n/2)}|\mathbf{X}|^{(p-\nu-1)/2}\exp[-\text{trace}(\boldsymbol{\Omega}^{-1}\mathbf{X})/2]$

Mean: $\text{E}(X_{jk}) = \nu\Omega_{jk}$

Variance: $\text{V}(X_{jk}) = \nu(\Omega_{jk}^2 + \Omega_{jj}\Omega_{kk})$

Marginal distributions: $X_{jj} \sim \text{Gamma}(\nu/2, \Omega_{jj}/2)$

Notes: If ν is an integer and $\mathbf{Z}_1, ..., \mathbf{Z}_\nu \overset{iid}{\sim} \text{Normal}(\mathbf{0}, \boldsymbol{\Omega})$, then $\sum_{i=1}^{\nu} \mathbf{Z}_i\mathbf{Z}_i^T \sim$ Wishart$(\nu, \boldsymbol{\Omega})$.

Inverse Wishart

Notation: $\mathbf{X} \sim \text{InvWishart}(\nu, \boldsymbol{\Omega})$

Support: $\mathbf{X} = \{X_{jk}\}$ is a $p \times p$ symmetric positive definite matrix

Parameters: degrees of freedom $\nu > p-1$ and $p \times p$ symmetric positive definite matrix $\boldsymbol{\Omega} = \{\Omega_{jk}\}$

PDF: $\frac{|\boldsymbol{\Omega}|^{\nu/2}}{2^{p\nu}\Gamma_p(n/2)}|\mathbf{X}|^{-(p-\nu-1)/2}\exp[-\text{trace}(\boldsymbol{\Omega}\mathbf{X}^{-1})/2]$

Mean: $\text{E}(X_{jk}) = \frac{1}{\nu-p-1}\Omega_{jk}$ (for $\nu > p+1$)

Variance: $\text{V}(X_{jk}) = \frac{(\nu-p+1)\Omega_{kk}^2 + (\nu-p-1)\Omega_{jj}\Omega_{kk}}{(\nu-p)(\nu-p-1)^2(\nu-p-3)}$ (for $\nu > p+3$)

Marginal distributions: $X_{jj} \sim \text{InvGamma}((\nu-p+1)/2, \Omega_{jj}/2)$

Notes: If $\mathbf{Y} \sim \text{Wishart}(\nu, \boldsymbol{\Omega}^{-1})$ then $\mathbf{Y}^{-1} \sim \text{InvWishart}(\nu, \boldsymbol{\Omega})$; if $\nu = p+1$ and $\boldsymbol{\Omega}$ is a diagonal matrix then the correlation $X_{jk}/\sqrt{X_{jj}X_{kk}} \sim \text{Uniform}(-1, 1)$.

A.2: List of conjugacy pairs

Below is a partial list of conjugacy pairs. In these derivations, all parameters not assigned a prior are assumed to be fixed.

1. **Binomial proportion**

   Likelihood: $Y|\theta \sim \text{Binomial}(n, \theta)$
   Prior: $\theta \sim \text{Beta}(a, b)$
   Posterior: $\theta|Y \sim \text{Beta}(a + Y, b + n - Y)$

2. **Negative-binomial proportion**

   Likelihood: $Y|\theta \sim \text{NegBinomial}(\theta, m)$
   Prior: $\theta \sim \text{Beta}(a, b)$
   Posterior: $\theta|Y \sim \text{Beta}(a + m, b + Y)$

3. **Multinomial probabilities**

   Likelihood: $\mathbf{Y} = (Y_1, ..., Y_p)|\boldsymbol{\theta} \sim \text{Multinomial}(n, \boldsymbol{\theta})$
   Prior: $\boldsymbol{\theta} \sim \text{Dirichlet}(\boldsymbol{\alpha})$ with $\boldsymbol{\alpha} = (\alpha_1, ..., \alpha_p)$
   Posterior: $\boldsymbol{\theta}|\mathbf{Y} \sim \text{Dirichlet}(\boldsymbol{\alpha} + \mathbf{Y})$

4. **Poisson rate**

   Likelihood: $Y_1, ..., Y_n|\lambda \stackrel{indep}{\sim} \text{Poisson}(N_i\lambda)$ with N_i fixed
   Prior: $\lambda \sim \text{Gamma}(a, b)$
   Posterior: $\lambda|\mathbf{Y} \sim \text{Gamma}(a + \sum_{i=1}^{n} Y_i, b + \sum_{i=1}^{n} N_i)$

5. **Mean of a normal distribution**

   Likelihood: $Y_1, ..., Y_n|\mu \stackrel{iid}{\sim} \text{Normal}(\mu, \sigma^2)$
   Prior: $\mu \sim \text{Normal}(\theta, \sigma^2/m)$
   Posterior: $\mu|\mathbf{Y} \sim \text{Normal}\left(\frac{n\bar{Y}+m\theta}{n+m}, \frac{\sigma^2}{n+m}\right)$ for $\bar{Y} = \sum_{i=1}^{n} Y_i/n$

6. **Variance of a normal distribution**

   Likelihood: $Y_1, ..., Y_n|\sigma^2 \stackrel{indep}{\sim} \text{Normal}(\mu_i, \sigma^2)$
   Prior: $\sigma^2 \sim \text{InvGamma}(a, b)$
   Posterior: $\sigma^2|\mathbf{Y} \sim \text{InvGamma}(a + n/2, b + \sum_{i=1}^{n}(Y_i - \mu_i)^2/2)$

7. **Precision of a normal distribution**

   Likelihood: $Y_1, ..., Y_n|\tau^2 \stackrel{indep}{\sim} \text{Normal}(\mu_i, 1/\tau^2)$
   Prior: $\tau^2 \sim \text{Gamma}(a, b)$
   Posterior: $\tau^2|\mathbf{Y} \sim \text{Gamma}(a + n/2, b + \sum_{i=1}^{n}(Y_i - \mu_i)^2/2)$

8. **Mean vector of a multivariate normal distribution**

   Likelihood: $\mathbf{Y}_1, ..., \mathbf{Y}_n|\boldsymbol{\mu} \stackrel{indep}{\sim} \text{Normal}(\mathbf{X}_i\boldsymbol{\mu}, \Sigma_i)$
   Prior: $\boldsymbol{\mu} \sim \text{Normal}(\boldsymbol{\theta}, \boldsymbol{\Omega})$
   Posterior: $\boldsymbol{\mu}|\mathbf{Y} \sim \text{Normal}(\mathbf{VM}, \mathbf{V})$ with $\mathbf{V} = (\sum_{i=1}^{n} \mathbf{X}_i^T \Sigma_i^{-1} \mathbf{X}_i + \boldsymbol{\Omega}^{-1})^{-1}$ and $\mathbf{M} = \sum_{i=1}^{n} \mathbf{X}_i^T \Sigma_i^{-1} \mathbf{Y}_i + \boldsymbol{\Omega}^{-1}\boldsymbol{\theta}$

Special case: If $\mathbf{X}_i = \mathbf{I}$ and $\boldsymbol{\Sigma}_i = \boldsymbol{\Sigma}$ for all i, then $\mathbf{V} = (n\boldsymbol{\Sigma}^{-1} + \boldsymbol{\Omega}^{-1})^{-1}$ and $\mathbf{M} = n\boldsymbol{\Sigma}^{-1}\bar{\mathbf{Y}} + \boldsymbol{\Omega}^{-1}\boldsymbol{\theta}$

9. **Covariance matrix of a multivariate normal distribution**

   Likelihood: $\mathbf{Y}_1, ..., \mathbf{Y}_n | \boldsymbol{\Sigma} \overset{indep}{\sim} \text{Normal}(\boldsymbol{\mu}_i, \boldsymbol{\Sigma})$

   Prior: $\boldsymbol{\Sigma} \sim \text{InvWishart}(\nu, \mathbf{R})$

   Posterior: $\boldsymbol{\Sigma}|\mathbf{Y} \sim \text{InvWishart}(n + \nu, \mathbf{S} + \mathbf{R})$, where $\mathbf{S} = \sum_{i=1}^n (\mathbf{Y}_i - \boldsymbol{\mu}_i)(\mathbf{Y}_i - \boldsymbol{\mu}_i)^T$

10. **Precision matrix of a multivariate normal distribution**

    Likelihood: $\mathbf{Y}_1, ..., \mathbf{Y}_n | \boldsymbol{\Omega} \overset{indep}{\sim} \text{Normal}(\boldsymbol{\mu}_i, \boldsymbol{\Omega}^{-1})$

    Prior: $\boldsymbol{\Omega} \sim \text{Wishart}(\nu, \mathbf{R})$

    Posterior: $\boldsymbol{\Omega}|\mathbf{Y} \sim \text{Wishart}\left(n + \nu, \left[\mathbf{S} + \mathbf{R}^{-1}\right]^{-1}\right)$, where $\mathbf{S} = \sum_{i=1}^n (\mathbf{Y}_i - \boldsymbol{\mu}_i)(\mathbf{Y}_i - \boldsymbol{\mu}_i)^T$

11. **Scale parameter of a gamma distribution**

    Likelihood: $Y_1, ..., Y_n | \mu \overset{iid}{\sim} \text{Gamma}(a_i, w_i b)$

    Prior: $b \sim \text{Gamma}(u, v)$

    Posterior: $b|\mathbf{Y} \sim \text{Gamma}(\sum_{i=1}^n a_i + u, \sum_{i=1}^n w_i Y_i + v)$

12. **Arbitrary parameter with discrete prior**

    Likelihood: $\mathbf{Y}_1, ..., \mathbf{Y}_n | \boldsymbol{\theta} \overset{indep}{\sim} f_i(\mathbf{Y}_i | \boldsymbol{\theta})$

    Prior: $\text{Prob}(\boldsymbol{\theta} = \boldsymbol{\theta}_k) = \pi_k$ for $\boldsymbol{\theta} \in \{\boldsymbol{\theta}_1, ..., \boldsymbol{\theta}_m\}$

    Posterior: $\text{Prob}(\boldsymbol{\theta} = \boldsymbol{\theta}_k | \mathbf{Y}) = L_k / [\sum_{j=1}^m L_j]$ where $L_k = \pi_k \prod_{i=1}^n f_i(\mathbf{Y}_i | \boldsymbol{\theta}_k)$

A.3: Derivations

Normal-normal model for a mean

Say $Y_i|\mu \overset{iid}{\sim} \text{Normal}(\mu, \sigma^2)$ for $i = 1,...,n$ with σ^2 known and prior $\mu \sim \text{Normal}(\theta, \sigma^2/m)$. Since the $Y_1, ..., Y_n$ are independent, the likelihood factors as

$$f(\mathbf{Y}|\mu) = \prod_{i=1}^{n} f(Y_i|\mu) = \prod_{i=1}^{n} \frac{1}{\sqrt{2\pi\sigma}} \exp\left[-\frac{(Y_i - \mu)^2}{2\sigma^2}\right].$$

Discarding constants that do not depend on μ and expressing the product of exponentials as the exponential of the sum, the likelihood is

$$f(\mathbf{Y}|\mu) \propto \exp\left[-\sum_{i=1}^{n} \frac{(Y_i - \mu)^2}{2\sigma^2}\right] \propto \exp\left[-\frac{1}{2}\left(-2\frac{n\bar{Y}}{\sigma^2}\mu + \frac{n}{\sigma^2}\mu^2\right)\right]$$

where $\bar{Y} = \sum_{i=1}^{n} Y_i/n$. The last equality comes from multiplying the quadratic terms, collecting them as a function of their power of μ, and discarding terms without a μ. Similarly, the prior can be written

$$\pi(\mu) \propto \exp\left[-\frac{m(\mu - \theta)^2}{2\sigma^2}\right] \propto \exp\left[-\frac{1}{2}\left(-2\frac{m\theta}{\sigma^2}\mu + \frac{m}{\sigma^2}\mu^2\right)\right].$$

Because both the likelihood and prior are quadratic in μ they can be combined as

$$
\begin{aligned}
p(\mu|\mathbf{Y}) &\propto f(\mathbf{Y}|\mu)\pi(\mu) \\
&\propto \exp\left[-\frac{1}{2}\left(-2\frac{n\bar{Y} + m\theta}{\sigma^2}\mu + \frac{n + m}{\sigma^2}\mu^2\right)\right] \\
&\propto \exp\left[-\frac{1}{2}\left(-2M\mu + \frac{1}{V}\mu^2\right)\right],
\end{aligned}
$$

where $M = (n\bar{Y} + m\theta)/\sigma^2$ and $V = \sigma^2/(n+m)$. The exponent of the posterior is quadratic in μ, and we have seen that a Gaussian PDF is quadratic in the exponent. Therefore, we rearrange the terms in the posterior to reveal its Gaussian PDF form. Completing the square in the exponent (and discarding and/or adding terms that do not depend on μ) gives

$$p(\mu|\mathbf{Y}) \propto \exp\left[-\frac{1}{2}\left(-2M\mu + \frac{1}{V}\mu^2\right)\right] \propto \exp\left[-\frac{(\mu - VM)^2}{2V}\right].$$

Therefore, the posterior is $\mu|\mathbf{Y} \sim N(VM, V)$. Plugging in the above expressions for M and V gives

$$\mu|\mathbf{Y} \sim N\left(w\bar{Y} + (1 - w)\theta, \frac{\sigma^2}{n + m}\right)$$

where $w = n/(n + m)$.

Normal-normal model for a mean vector

The model is

$$\mathbf{Y}|\boldsymbol{\beta} \sim \text{Normal}(\mathbf{X}\boldsymbol{\beta}, \boldsymbol{\Sigma}) \text{ and } \boldsymbol{\beta} \sim \text{Normal}(\boldsymbol{\mu}, \boldsymbol{\Omega}).$$

As with the normal-normal model in Section 7.4, we proceed by expressing the exponential of the posterior as a quadratic form in $\boldsymbol{\beta}$ and then comparing this expression to a multivariate normal to determine the posterior. Using precision matrices $\mathbf{U} = \boldsymbol{\Sigma}^{-1}$ and $\mathbf{V} = \boldsymbol{\Omega}^{-1}$, the posterior is

$$
\begin{aligned}
p(\boldsymbol{\beta}|\mathbf{Y}) &\propto f(\mathbf{Y}|\boldsymbol{\beta})\pi(\boldsymbol{\beta}) \\
&\propto \exp\left[-\frac{1}{2}(\mathbf{Y} - \mathbf{X}\boldsymbol{\beta})^T\mathbf{U}(\mathbf{Y} - \mathbf{X}\boldsymbol{\beta})^T\right]\left[-\frac{1}{2}(\boldsymbol{\beta} - \boldsymbol{\mu})^T\mathbf{V}(\boldsymbol{\beta} - \boldsymbol{\mu})^T\right] \\
&\propto \exp\left\{-\frac{1}{2}\left[-2(\mathbf{Y}^T\mathbf{U}\mathbf{X} + \boldsymbol{\mu}^T\mathbf{V})\boldsymbol{\beta} + \boldsymbol{\beta}(\mathbf{X}^T\mathbf{U}\mathbf{X} + \mathbf{V})\boldsymbol{\beta}\right]\right\} \\
&\propto \exp\left[-\frac{1}{2}\left(-2\mathbf{W}^T\boldsymbol{\beta} + \boldsymbol{\beta}^T\mathbf{P}\boldsymbol{\beta}\right)\right]
\end{aligned}
$$

where $\mathbf{W} = \mathbf{X}^T\mathbf{U}\mathbf{Y} + \mathbf{V}\boldsymbol{\mu}$ and $\mathbf{P} = \mathbf{X}^T\mathbf{U}\mathbf{X} + \mathbf{V}$. If $\boldsymbol{\beta}|\mathbf{Y} \sim \text{Normal}(\mathbf{M}, \mathbf{S})$ for some mean vector $\mathbf{M}$ and covariance matrix $\mathbf{S}$, then its PDF can be written

$$
\begin{aligned}
p(\boldsymbol{\beta}|\mathbf{Y}) &\propto \exp\left[-\frac{1}{2}(\boldsymbol{\beta} - \mathbf{M})^T\mathbf{S}^{-1}(\boldsymbol{\beta} - \mathbf{M})\right] \\
&\propto \exp\left[-\frac{1}{2}\left(-2\mathbf{M}^T\mathbf{S}^{-1}\boldsymbol{\beta} + \boldsymbol{\beta}^T\mathbf{S}^{-1}\boldsymbol{\beta}\right)\right].
\end{aligned}
$$

To reconcile these two expressions of the posterior we must have posterior covariance $\mathbf{S} = \mathbf{P}^{-1}$ and $\mathbf{S}^{-1}\mathbf{M} = \mathbf{W}$ and thus $\mathbf{M} = \mathbf{S}\mathbf{W} = \mathbf{P}^{-1}\mathbf{W}$. Inserting the expressions for $\mathbf{W}$ and $\mathbf{P}$ and replacing precision matrices with covariance matrices gives the posterior

$$\boldsymbol{\beta}|\mathbf{Y} \sim \text{Normal}\left[\boldsymbol{\Sigma}_\beta(\mathbf{X}^T\boldsymbol{\Sigma}^{-1}\mathbf{Y} + \boldsymbol{\Omega}^{-1}\boldsymbol{\mu}), \boldsymbol{\Sigma}_\beta\right],$$

where $\boldsymbol{\Sigma}_\beta = (\mathbf{X}'\boldsymbol{\Sigma}^{-1}\mathbf{X} + \boldsymbol{\Omega}^{-1})^{-1}$

Normal-inverse Wishart model for a covariance matrix

The model for the p-vectors $\mathbf{Y}_1, ..., \mathbf{Y}_n$ given the $p \times p$ covariance matrix $\boldsymbol{\Sigma}$ is

$$\mathbf{Y}_i \stackrel{indep}{\sim} \text{Normal}(\boldsymbol{\mu}_i, \boldsymbol{\Sigma}) \text{ and } \boldsymbol{\Sigma} \sim \text{InvWishart}_p(\nu, \mathbf{R}).$$

Using the facts that for arbitrary matrices $\mathbf{A}$, $\mathbf{B}$ and $\mathbf{C}$, $\text{Trace}(\mathbf{A} + \mathbf{B}) = \text{Trace}(\mathbf{A}) + \text{Trace}(\mathbf{B})$ and $\text{Trace}(\mathbf{A}\mathbf{B}\mathbf{C}) = \text{Trace}(\mathbf{B}\mathbf{C}\mathbf{A})$, the likelihood can be written

$$f(\mathbf{Y}|\boldsymbol{\Sigma}) \quad \propto \quad \prod_{i=1}^{n} |\boldsymbol{\Sigma}|^{-1/2} \exp\left[-\frac{1}{2}(\mathbf{Y}_i - \boldsymbol{\mu}_i)^T \boldsymbol{\Sigma}^{-1}(\mathbf{Y}_i - \boldsymbol{\mu}_i)\right]$$

$$\propto \quad |\boldsymbol{\Sigma}|^{-n/2} \exp\left[-\frac{1}{2}\sum_{i=1}^{n}(\mathbf{Y}_i - \boldsymbol{\mu}_i)^T \boldsymbol{\Sigma}^{-1}(\mathbf{Y}_i - \boldsymbol{\mu}_i)\right]$$

$$\propto \quad |\boldsymbol{\Sigma}|^{-n/2} \exp\left\{-\frac{1}{2}\sum_{i=1}^{n}\mathrm{Trace}\left[(\mathbf{Y}_i - \boldsymbol{\mu}_i)^T \boldsymbol{\Sigma}^{-1}(\mathbf{Y}_i - \boldsymbol{\mu}_i)\right]\right\}$$

$$\propto \quad |\boldsymbol{\Sigma}|^{-n/2} \exp\left\{-\frac{1}{2}\sum_{i=1}^{n}\mathrm{Trace}\left[\boldsymbol{\Sigma}^{-1}(\mathbf{Y}_i - \boldsymbol{\mu}_i)(\mathbf{Y}_i - \boldsymbol{\mu}_i)^T\right]\right\}$$

$$\propto \quad |\boldsymbol{\Sigma}|^{-n/2} \exp\left\{-\frac{1}{2}\mathrm{Trace}\left[\sum_{i=1}^{n}\boldsymbol{\Sigma}^{-1}(\mathbf{Y}_i - \boldsymbol{\mu}_i)(\mathbf{Y}_i - \boldsymbol{\mu}_i)^T\right]\right\}$$

$$\propto \quad |\boldsymbol{\Sigma}|^{-n/2} \exp\left\{-\frac{1}{2}\mathrm{Trace}\left[\boldsymbol{\Sigma}^{-1}\sum_{i=1}^{n}(\mathbf{Y}_i - \boldsymbol{\mu}_i)(\mathbf{Y}_i - \boldsymbol{\mu}_i)^T\right]\right\}$$

$$\propto \quad |\boldsymbol{\Sigma}|^{-n/2} \exp\left[-\frac{1}{2}\mathrm{Trace}(\boldsymbol{\Sigma}^{-1}\mathbf{W})\right]$$

where $\mathbf{W} = \sum_{i=1}^{n}(\mathbf{Y}_i - \boldsymbol{\mu}_i)(\mathbf{Y}_i - \boldsymbol{\mu}_i)^T$. The inverse Wishart prior is

$$\pi(\boldsymbol{\Sigma}) \propto |\boldsymbol{\Sigma}|^{-(\nu+p+1)/2} \exp\left[-\frac{1}{2}\mathrm{Trace}(\boldsymbol{\Sigma}^{-1}\mathbf{R})\right].$$

Combining the likelihood and prior, the posterior is

$$p(\boldsymbol{\Sigma}|\mathbf{Y}) \quad \propto \quad f(\mathbf{Y}|\boldsymbol{\Sigma})\pi(\boldsymbol{\Sigma})$$

$$\propto \quad |\boldsymbol{\Sigma}|^{-(n+\nu+p+1)/2} \exp\left\{-\frac{1}{2}\mathrm{Trace}[\boldsymbol{\Sigma}^{-1}(\mathbf{W} + \mathbf{R})]\right\}.$$

Therefore, $\boldsymbol{\Sigma}|\mathbf{Y} \sim \mathrm{InvWishart}_p(n + \nu, \sum_{i=1}^{n}(\mathbf{Y}_i - \boldsymbol{\mu}_i)(\mathbf{Y}_i - \boldsymbol{\mu}_i)^T + \mathbf{R})$.

Jeffreys' prior for a normal model

The Gaussian model is $Y_i \overset{iid}{\sim} \mathrm{Normal}(\mu, \sigma^2)$. Denote $\tau = \sigma^2$. The log likelihood is

$$\log f(\mathbf{Y}|\mu, \tau) = -\frac{n}{2}\log(\tau) - \frac{1}{2\tau}\sum_{i=1}^{n}(Y_i - \mu)^2.$$

The information matrix depends on both second derivatives and the cross derivative. The second derivatives are

$$\frac{\partial^2 \log f(\mathbf{Y}|\mu, \tau)}{\partial \mu^2} \quad = \quad \frac{\partial}{\partial \mu}\frac{1}{\tau}\sum_{i=1}^{n}(Y_i - \mu) = -\frac{n}{\tau}$$

and

$$\frac{\partial^2 \log f(\mathbf{Y}|\mu,\tau)}{\partial\tau^2} = \frac{\partial}{\partial\tau}\frac{-n}{2\tau} + \frac{1}{2\tau^2}\sum_{i=1}^{n}(Y_i - \mu)^2$$

$$= \frac{n}{2\tau^2} - \frac{1}{\tau^3}\sum_{i=1}^{n}(Y_i - \mu)^2.$$

The cross derivative is

$$\frac{\partial^2 \log f(\mathbf{Y}|\mu,\tau)}{\partial\mu\partial\tau} = \frac{\partial}{\partial\tau}\frac{1}{\tau}\sum_{i=1}^{n}(Y_i - \mu) = -\frac{1}{\tau^2}\sum_{i=1}^{n}(Y_i - \mu)$$

Since $E(Y_i) = \mu$ and $E(Y_i - \mu)^2 = \tau$, the elements of the information matrix are

$$-E\left(\frac{\partial^2 \log f(\mathbf{Y}|\mu,\tau)}{\partial\mu^2}\right) = \frac{n}{\tau}$$

$$-E\left(\frac{\partial^2 \log f(\mathbf{Y}|\mu,\tau)}{\partial\tau^2}\right) = -\frac{n}{2\tau^2} + \frac{n\tau}{\tau^3} = \frac{n}{2\tau^2}$$

$$-E\left(\frac{\partial^2 \log f(\mathbf{Y}|\mu,\tau)}{\partial\mu\partial\tau}\right) = 0.$$

The determinant of the 2×2 information matrix is thus

$$|I(\mu,\tau)| = \left(\frac{n}{\tau}\right)\left(\frac{n}{2\tau^2}\right) - 0^2 = \frac{n^2}{2\tau^3},$$

and the JP is

$$\pi(\mu,\sigma) \propto \sqrt{\frac{n^2}{2\tau^3}} \propto \frac{1}{(\sigma^2)^{3/2}}.$$

Jeffreys' prior for multiple linear regression

Assume $\mathbf{Y}|\boldsymbol{\beta},\sigma^2 \sim \text{Normal}(\mathbf{X}\boldsymbol{\beta}, \sigma^2\mathbf{I}_n)$ and denote $\tau = \sigma^2$. The log likelihood is

$$\log f(\mathbf{Y}|\boldsymbol{\beta},\tau) = -\frac{n}{2}\log(\tau) - \frac{1}{2\tau}(\mathbf{Y} - \mathbf{X}\boldsymbol{\beta})^T(\mathbf{Y} - \mathbf{X}\boldsymbol{\beta}).$$

The second derivative with respect to τ is

$$\frac{\partial^2 \log f(\mathbf{Y}|\boldsymbol{\beta},\tau)}{\partial\tau^2} = \frac{\partial}{\partial\tau}\frac{-n}{2\tau} + \frac{1}{2\tau^2}(\mathbf{Y} - \mathbf{X}\boldsymbol{\beta})^T(\mathbf{Y} - \mathbf{X}\boldsymbol{\beta})$$

$$= \frac{n}{2\tau^2} - \frac{1}{\tau^3}(\mathbf{Y} - \mathbf{X}\boldsymbol{\beta})^T(\mathbf{Y} - \mathbf{X}\boldsymbol{\beta}).$$

Taking derivatives with respect to $\boldsymbol{\beta}$ requires using matrix calculus identities including the formula for the derivative of a quadratic form,

$$\frac{\partial^2 \log f(\mathbf{Y}|\boldsymbol{\beta},\tau)}{\partial\boldsymbol{\beta}^2} = \frac{\partial}{\partial\boldsymbol{\beta}}\frac{1}{\tau}\mathbf{X}^T(\mathbf{Y} - \mathbf{X}\boldsymbol{\beta}) = -\frac{1}{\tau}\mathbf{X}^T\mathbf{X}.$$

The cross derivative is

$$\frac{\partial^2 \log f(\mathbf{Y}|\boldsymbol{\beta}, \tau)}{\partial\boldsymbol{\beta}\partial\tau} = \frac{\partial}{\partial\tau}\frac{1}{\tau}\mathbf{X}^T(\mathbf{Y}-\mathbf{X}\boldsymbol{\beta}) = -\frac{1}{\tau^2}\mathbf{X}^T(\mathbf{Y}-\mathbf{X}\boldsymbol{\beta}).$$

Since $\mathrm{E}(\mathbf{Y}) = \mathbf{X}\boldsymbol{\beta}$ and $\mathrm{E}(\mathbf{Y}-\mathbf{X}\boldsymbol{\beta})^T(\mathbf{Y}-\mathbf{X}\boldsymbol{\beta}) = n\tau$, the elements of the information matrix are

$$-\mathrm{E}\left(\frac{\partial^2 \log f(\mathbf{Y}|\mu,\tau)}{\partial\boldsymbol{\beta}^2}\right) = \frac{1}{\tau}\mathbf{X}^T\mathbf{X}$$

$$-\mathrm{E}\left(\frac{\partial^2 \log f(\mathbf{Y}|\mu,\tau)}{\partial\tau^2}\right) = -\frac{n}{2\tau^2}+\frac{n\tau}{\tau^3} = \frac{n}{2\tau^2}$$

$$-\mathrm{E}\left(\frac{\partial^2 \log f(\mathbf{Y}|\boldsymbol{\beta},\tau)}{\partial\mu\partial\tau}\right) = 0.$$

The determinant of the $(p+1)\times(p+1)$ block-diagonal information matrix is thus

$$|I(\boldsymbol{\beta},\tau)| = \left|\frac{1}{\tau}\mathbf{X}^T\mathbf{X}\right|\frac{n}{2\tau^2} = \frac{n}{2\tau^{p+2}}|\mathbf{X}^T\mathbf{X}|,$$

and the JP is

$$\pi(\boldsymbol{\beta},\sigma^2) \propto \sqrt{\frac{n}{2\tau^{p+2}}\left|\mathbf{X}^T\mathbf{X}\right|} \propto \frac{1}{(\sigma^2)^{p/2+1}}.$$

Convergence of the Gibbs sampler

Here we provide: (1) a proof that the Gibbs sampler generates posterior samples after convergence and (2) a discussion of the theory of Markov processes that ensures that Gibbs sampler converges to the posterior distribution.

Part (1): The proof of (1) is equivalent to showing that the posterior distribution is the stationary distribution of this Markov chain. That is, if we make a draw from the posterior distribution and then iterate the Gibbs sampler forward one iteration from this starting point, the next iteration also follows the posterior distribution. To make the derivations tractable, we restrict the proof to the bivariate case with $p = 2$ and thus $\boldsymbol{\theta} = (\theta_1, \theta_2)$ and denote the posterior density as $p(\theta_1, \theta_2) = p(\boldsymbol{\theta}|\mathbf{Y})$, the full conditional density as $p(\theta_1|\theta_2) = p(\theta_1|\theta_2, \mathbf{Y})$, and the marginal density as $p(\theta_1) = \int p(\theta_1, \theta_2)d\theta_2 = \int p(\theta_1|\theta_2)p(\theta_2)d\theta_2$. Assume we have reached convergence and so one draw in the chain is a realization from the posterior distribution, say $\boldsymbol{\theta}^* = (\theta_1^*, \theta_2^*) \sim p(\theta_1, \theta_2)$. We would like to show that the subsequent sample also follows the posterior distribution. By recursion, this shows that once the algorithm has converged, all samples follow the posterior.

The next sample, (θ_1', θ_2'), drawn from Gibbs sampling has density

$$q(\theta_1', \theta_2'|\theta_1^*, \theta_2^*) = p(\theta_1'|\theta_2^*)p(\theta_2'|\theta_1'),$$

where the two elements of the product represent the updates of the two parameters from their full conditional distribuitons given the current value of the parameters in the chain. We want to show that the marginal distribution of (θ'_1, θ'_2) integrating over (θ^*_1, θ^*_2) follows the posterior. The marginal distribution is

$$g(\theta'_1, \theta'_2) = \int \int q(\theta'_1, \theta'_2 | \theta^*_1, \theta^*_2) f(\theta^*_1, \theta^*_2) d\theta^*_1 d\theta^*_2.$$

The integral reduces to

$$
\begin{aligned}
g(\theta'_1, \theta'_2) &= \int \int p(\theta'_1 | \theta^*_2) p(\theta'_2 | \theta'_1) p(\theta^*_1 | \theta^*_2) p(\theta^*_2) d\theta^*_2 d\theta^*_1 \\
&= p(\theta'_2 | \theta'_1) \int p(\theta'_1 | \theta^*_2) p(\theta^*_2) \left[\int p(\theta^*_1 | \theta^*_2) d\theta^*_1 \right] d\theta^*_2 \\
&= p(\theta'_2 | \theta'_1) \int p(\theta'_1 | \theta^*_2) p(\theta^*_2) d\theta^*_2 \\
&= p(\theta'_2 | \theta'_1) \int p(\theta'_1, \theta^*_2) d\theta^*_2 \\
&= p(\theta'_2 | \theta'_1) p(\theta'_1) = p(\theta'_1, \theta'_2),
\end{aligned}
$$

as desired. The proof for $p > 2$ similar but involves higher-order integration.

Part (2): Part (1) shows (for a special case) that the stationary distribution of the Gibbs sampler is the posterior distribution. The proof that the Gibbs sampler converges to its stationary (posterior) distribution draws heavily from Markov chain theory. Given that the posterior distribution is the stationary distribution, [82] proves that a Gibbs sampler converges to the posterior distribution if the chain is aperiodic and p-irreducible. A chain is *aperiodic* if for any partition of the posterior domain of $\boldsymbol{\theta}$, say $\{\mathcal{A}_1, ..., \mathcal{A}_m\}$, so that each subset has positive posterior probability, then the probability of the chain transitioning from $\mathcal{A}_i$ to $\mathcal{A}_j$ is positive for any i and j. A chain is *p-irreducible* if for any initial value $\boldsymbol{\theta}^{(0)}$ in the support of the posterior distribution and set $\mathcal{A}$ with positive posterior probability, i.e., $\text{Prob}(\boldsymbol{\theta} \in \mathcal{A}) > 0$, there exists an s so that there is a positive probability that the chain will visit $\mathcal{A}$ at iteration s. Proving convergence then requires showing that the Gibbs sampler is aperiodic and p-irreducible which is discussed, e.g., in [82] and [69]. A sufficient condition is that for any set $\mathcal{A}$ with positive posterior probability and any initial value $\boldsymbol{\theta}^{(0)}$ in the support of the posterior distribution, the probability under the Gibbs sampler that $\boldsymbol{\theta}^{(1)} \in \mathcal{A}$ is positive. This condition is met in all but exotic cases where support of full conditional distributions depend on the values of other parameters, in which case convergence should be studied carefully.

Marginal distribution of a normal mean under Jeffreys' prior

Assume $Y_i \stackrel{iid}{\sim} \text{Normal}(\mu, \sigma^2)$ and Jeffreys' prior $\pi(\mu, \sigma^2) \propto (\sigma^2)^{-3/2}$. Denoting $\tau = \sigma^2$, the joint posterior is

$$p(\mu, \tau | \mathbf{Y}) \quad \propto \quad \left\{ \tau^{-n/2} \exp\left[-\frac{\sum_{i=1}^n (Y_i - \mu)^2}{2\tau} \right] \right\} \left\{ \tau^{-3/2} \right\}$$

$$\propto \quad \tau^{-(n+1)/2-1} \exp\left[-\frac{\sum_{i=1}^n (Y_i - \mu)^2}{2\tau} \right]$$

$$\propto \quad \tau^{-A-1} \exp\left[-\frac{B}{\tau} \right],$$

where $A = (n+1)/2$ and $B = \sum_{i=1}^n (Y_i - \mu)^2 / 2$. As a function of τ, the joint distribution resembles an $\text{InvGamma}(A, B)$ PDF. Integrating over τ gives

$$p(\mu | \mathbf{Y}) \quad \propto \quad \int p(\mu, \tau | bY) d\tau$$

$$\propto \quad \int \tau^{-A-1} \exp(-B/\tau) d\tau$$

$$\propto \quad \frac{\Gamma(A)}{B^A} \int \frac{B^A}{\Gamma(A)} \tau^{-A-1} \exp(-B/\tau) d\tau$$

$$\propto \quad \frac{\Gamma(A)}{B^A}$$

$$\propto \quad \left[\sum_{i=1}^n (Y_i - \mu)^2 \right]^{-(n+1)/2} .$$

The marginal PDF is a quadratic function of μ raised to the power $-(n+1)/2$, suggesting that the posterior is a t distribution with n degrees of freedom. Completing the square gives

$$\sum_{i=1}^n (Y_i - \mu)^2 \quad = \quad \sum_{i=1}^n Y_i^2 - 2 \sum_{i=1}^n Y_i \mu + n\mu^2$$

$$= \quad n \left[\sum_{i=1}^n Y_i^2 / n - 2\bar{Y}\mu + \mu^2 \right]$$

$$= \quad n \left[\sum_{i=1}^n Y_i^2 / n - \bar{Y}^2 + \bar{Y}^2 - 2\bar{Y}\mu + \mu^2 \right]$$

$$= \quad n \left[\sum_{i=1}^n Y_i^2 / n - \bar{Y}^2 + (\mu - \bar{Y})^2 \right]$$

$$= \quad n \left[\hat{\sigma}^2 + (\mu - \bar{Y})^2 \right],$$

since $\hat{\sigma}^2 = \sum_{i=1}^n (Y_i - \bar{Y})^2 / n = \sum_{i=1}^n Y_i^2 / n - \bar{Y}^2$. Inserting this expression back into the marginal posterior gives

$$p(\mu|\mathbf{Y}) \quad \propto \quad \left[\sum_{i=1}^{n}(Y_i - \mu)^2\right]^{-(n+1)/2}$$

$$\propto \quad \left[\hat{\sigma}^2 + (\mu - \bar{Y})^2\right]^{-(n+1)/2}$$

$$\propto \quad \left[1 + \frac{1}{n}\left(\frac{\mu - \bar{Y}}{\hat{\sigma}/\sqrt{n}}\right)^2\right]^{-(n+1)/2}.$$

This is Student's t distribution with location parameter $\bar{Y}$, scale parameter $\hat{\sigma}/\sqrt{n}$, and n degrees of freedom.

Marginal posterior of the regression coefficients under Jeffreys' prior

Assume $\mathbf{Y}|\beta, \sigma^2 \sim \text{Normal}(\mathbf{X}\beta, \sigma^2\mathbf{I}_n)$ and Jeffreys' prior $\pi(\beta, \sigma^2) \propto (\sigma^2)^{-p/2-1}$. Denoting $\tau = \sigma^2$, the joint posterior is

$$p(\beta, \tau|\mathbf{Y}) \quad \propto \quad \left\{\tau^{-n/2}\exp\left[-\frac{1}{2\tau}(\mathbf{Y} - \mathbf{X}\beta)^T(\mathbf{Y} - \mathbf{X}\beta)\right]\right\}\tau^{-p/2-1}$$

$$\propto \quad \tau^{-A-1}\exp\left[-\frac{B}{\tau}\right],$$

where $A = (n+p)/2$ and $B = (\mathbf{Y} - \mathbf{X}\beta)^T(\mathbf{Y} - \mathbf{X}\beta)/2$. Marginalizing over σ^2 gives

$$p(\beta|\mathbf{Y}) \quad = \quad \int p(\beta, \tau|\mathbf{Y})d\tau$$

$$\propto \quad \frac{\Gamma(A)}{B^A}\int\frac{B^A}{\Gamma(A)}\tau^{-A-1}\exp\left[-\frac{B}{\tau}\right]d\tau$$

$$\propto \quad B^{-A}$$

$$\propto \quad \left[(\mathbf{Y} - \mathbf{X}\beta)^T(\mathbf{Y} - \mathbf{X}\beta)\right]^{-(n+p)/2}.$$

The quadratic form is factored as

$$(\mathbf{Y} - \mathbf{X}\beta)^T(\mathbf{Y} - \mathbf{X}\beta) \quad = \quad \mathbf{Y}^T\mathbf{Y} - 2\mathbf{Y}^T\mathbf{X}\beta + \beta^T\mathbf{W}\beta$$

$$= \quad \mathbf{Y}^T\mathbf{Y} - 2\hat{\beta}^T\mathbf{W}\beta + \beta^T\mathbf{W}\beta$$

$$= \quad \mathbf{Y}^T\mathbf{Y} - \hat{\beta}^T\mathbf{W}\hat{\beta} + \hat{\beta}^T\mathbf{W}\hat{\beta} - 2\hat{\beta}^T\mathbf{W}\beta + \beta^T\mathbf{W}\beta$$

$$= \quad n\hat{\sigma}^2 + (\beta - \hat{\beta})^T\mathbf{W}(\beta - \hat{\beta})$$

where $\mathbf{W} = \mathbf{X}^T\mathbf{X}$, $\hat{\beta} = (\mathbf{W})^{-1}\mathbf{X}^T\mathbf{Y}$ is the usual least squares estimator, and $n\hat{\sigma}^2 = (\mathbf{Y} - \mathbf{X}\hat{\beta})^T(\mathbf{Y} - \mathbf{X}\hat{\beta}) = \mathbf{Y}^T\mathbf{Y} - \hat{\beta}^T\mathbf{W}\hat{\beta}$. Therefore,

$$
\begin{aligned}
p(\beta|\mathbf{Y}) &\propto [(\mathbf{Y} - \mathbf{X}\beta)^T(\mathbf{Y} - \mathbf{X}\beta)]^{-(n+p)/2} \\
&\propto \left[n\hat{\sigma}^2 + (\beta - \hat{\beta})^T\mathbf{W}(\beta - \hat{\beta})\right]^{-(n+p)/2} \\
&\propto \left[1 + \frac{1}{n\hat{\sigma}^2}(\beta - \hat{\beta})^T\mathbf{W}(\beta - \hat{\beta})\right]^{-(n+p)/2}.
\end{aligned}
$$

The marginal posterior of β is thus the p-dimensional t-distribution with location vector $\hat{\beta}$, scale matrix $\hat{\sigma}^2(\mathbf{X}^T\mathbf{X})^{-1}$, and n degrees of freedom.

The two-sample t-test in Section 4.1.2 is a special case. If we parameterize the means as μ for the first group and $\mu + \delta$ for the second group then $\mathbf{X}$'s first column has all ones and its second column is n_1 zeros followed by n_2 ones, and $\beta = (\mu, \delta)^T$. This gives the least squares estimator as $\hat{\beta} = (\bar{Y}, \bar{Y}_2 - \bar{Y}_1)^T$ and $\hat{\sigma}^2 = (n_1\hat{\sigma}_1^2 + n_2\hat{\sigma}_2^2)/(n_1 + n_2)$ and

$$
(\mathbf{X}^T\mathbf{X})^{-1} = \begin{bmatrix} n_1 + n_2 & n_2 \\ n_2 & n_2 \end{bmatrix}^{-1} = \frac{1}{n_1 n_2}\begin{bmatrix} n_2 & -n_2 \\ -n_2 & n_1 + n_2 \end{bmatrix} = \begin{bmatrix} \frac{1}{n_1} & -\frac{1}{n_1} \\ -\frac{1}{n_1} & \frac{1}{n_1} + \frac{1}{n_2} \end{bmatrix}.
$$

This give the components of the joint posterior distribution of β, since the marginal distributions of multivariate t are univariate, we have $\delta|\mathbf{Y} \sim t_n\left[\bar{Y}_2 - \bar{Y}_1, \hat{\sigma}\left(\frac{1}{n_1} + \frac{1}{n_2}\right)\right]$

Proof of posterior consistency

Here we prove posterior consistency in the general but simple case with independent data and parameter with discrete support. Assume that:

(A1) $Y_i \overset{iid}{\sim} f(y|\theta)$ for $i = 1, ..., n$

(A2) The support is discrete, $\theta \in \{\theta_1, \theta_2, ...\} = \mathcal{S}$

(A3) The true value $\theta_0 \in \mathcal{S}$ has positive prior probability, $\pi(\theta_0) > 0$

(A4) The Kullback-Leibler divergence

$$
KL(\theta) = \mathrm{E}_{Y|\theta_0}\left[\log\left(\frac{f(Y|\theta_0)}{f(Y|\theta)}\right)\right]
$$

satisfies $KL(\theta) > 0$ for all $\theta \neq \theta_0$.

Assumption (A4) ensures that the parameter is identifiable by asserting that on average the likelihood is higher for true value and any other value.

Theorem 1 *Assuming (A1)-(A4), Prob$(\theta = \theta_0|Y_1, ..., Y_n) \to 1$ as $n \to \infty$.*

Proof 1 *For any $\theta \in \mathcal{S}$,*

$$\log\left[\frac{p(\theta|\mathbf{Y})}{p(\theta_0|\mathbf{Y})}\right] = \log\left[\frac{\pi(\theta)}{\pi(\theta_0)}\right] + \sum_{i=1}^{n} \log\left[\frac{f(Y_i|\theta)}{f(Y_i|\theta_0)}\right].$$

By the law of large numbers, $\frac{1}{n}\sum_{i=1}^{n}\log\left[\frac{f(Y_i|\theta)}{f(Y_i|\theta_0)}\right] \to -KL(\theta)$, and thus

$$\log\left[\frac{p(\theta|\mathbf{Y})}{p(\theta_0|\mathbf{Y})}\right] \to \log\left[\frac{\pi(\theta)}{\pi(\theta_0)}\right] - nKL(\theta).$$

Therefore, as $n \to \infty$, $p(\theta|\mathbf{Y})/p(\theta_0|\mathbf{Y}) \to 0$ for any $\theta \neq \theta_0$, and $Prob(\theta = \theta_0|\mathbf{Y})$ converges to one.

This proof can be generalized to continuous parameters by discretizng the support and making additional assumptions about the smoothness of the likelihood and prior density functions.

A.4: Computational algorithms

Integrated nested Laplace approximation (INLA)

INLA [73] is a deterministic approximation to the marginal posterior of each parameter that combines many of the ideas discussed in Section 3.1. The method is most fitting in the special but common case where the parameter vector $\theta = (\alpha, \beta)$ can be divided into a low-dimensional α and a high-dimensional β whose posterior is approximately Gaussian. For example, in a random effects model (Section 4.4) α might include the variance components and β might include all of the Gaussian random effects.

Evoking the Bayesian CLT (i.e., Laplace approximation) in Section 3.1.3, assume that the conditional posterior of β conditioned on α is approximately

$$\beta|\alpha, \mathbf{Y} \sim \text{Normal}(\mu(\alpha), \Sigma(\alpha))$$

and denote the corresponding density function as $\phi(\beta; \mu(\alpha), \Sigma(\alpha))$. We first use this approximation for the marginal distribution of the low-dimensional parameter α. Since $p(\alpha, \beta|\mathbf{Y}) = p(\beta|\alpha, \mathbf{Y})p(\alpha|\mathbf{Y})$, the marginal posterior of α can be written

$$p(\alpha|\mathbf{Y}) = \frac{p(\alpha, \beta|\mathbf{Y})}{p(\beta|\alpha, \mathbf{Y})}.$$

Expanding around the MAP estimate $\beta = \mu(\alpha)$ and using the Laplace approximation for the denominator gives the approximation

$$p(\alpha|\mathbf{Y}) \approx \frac{f(\mathbf{Y}|\alpha, \beta)\pi(\alpha, \beta|\mathbf{Y})}{\phi(\beta; \mu(\alpha), \Sigma(\alpha))}\Bigg|_{\beta=\mu(\alpha)}.$$

This low-dimensional distribution and can be evaluated using the methods in Section 3.1, e.g., grid approximations or numerical integration.

The Laplace approximation can also be used to approximate the marginal distribution of each element of β. Let β_{-i} be the elements of β excluding β_i. Following arguments similar to the approximation of the posterior of α,

$$p(\beta_i | \alpha, \mathbf{Y}) \propto \frac{f(\mathbf{Y} | \alpha, \beta) \pi(\alpha, \beta)}{p(\beta_{-i} | \beta_i, \alpha, \mathbf{Y})}.$$

This can be approximated using a Laplace approximation for $p(\beta_{-i} | \beta_i, \alpha, \mathbf{Y})$ around its posterior mode ([73] also consider faster approximations). Finally, to obtain $p(\beta_{-i} | \mathbf{Y})$ requires numerical integration over α, and therefore the Laplace approximation is nested within numerical integration.

Metropolis–adjusted Langevin algorithm

Metropolis–Hastings sampling (Section 3.2.2) is a flexible algorithm but depends on finding a reasonable candidate distribution. A Gaussian random walk distribution for the candidate $\theta^* = (\theta_1^*, ..., \theta_p^*)$ given the current value at the onset of iteration s, $\theta^{(s-1)}$, is

$$\theta^* | \theta^{(s-1)} \sim \text{Normal}(\theta^{(s-1)}, c^2 \mathbf{I}_p),$$

where $c > 0$ is a tuning parameter. This candidate is easy to code, very general and surprisingly effective. However, convergence can be improved by tailoring the candidate distribution to the problem at hand. We saw in Section 3.2 that if the candidate distribution is taken to be the full conditional distribution Metropolis–Hastings sampling becomes Gibbs sampling. While Gibbs sampling is free from tuning parameters and is thus easier to implement, it requires derivation of full conditional distributions which can be tedious and is not always possible.

The Metropolis-adjusted Langevin (MALA) algorithm [68] balances the strengths of random-walk Metropolis and Gibbs sampling. Rather than simply centering the candidate distribution on the current value, MALA uses the gradient of the posterior to push the candidate distribution towards the center of the distribution. This requires computing the gradient of the posterior and thus the algorithm is more complex than a random walk, but the gradient is typically easier to derive and more generally available than the full conditional distribution required for Gibbs sampling.

Define the gradient vector of the log posterior as $\nabla(\theta) = [\nabla_1(\theta), ..., \nabla_p(\theta)]^T$ where

$$\nabla_j(\theta) = \frac{\partial}{\partial \theta_j} \{\log[f(\mathbf{Y} | \theta)] + \log[\pi(\theta)]\}$$

is the partial derivative with respect to the jth parameter. The candidate is

$$\theta^* | \theta^{(s-1)} \sim \text{Normal}\left(\theta^{(s-1)} + \frac{c^2}{2} \nabla(\theta^{(s-1)}), c^2 \mathbf{I}_p\right).$$

Unlike the random-walk candidate distribution, the MALA candidate distribution is asymmetric and requires including the candidate distribution in the acceptance ratio,

$$R = \frac{f(\mathbf{Y}|\boldsymbol{\theta}^*)\pi(\boldsymbol{\theta}^*)}{f(\mathbf{Y}|\boldsymbol{\theta}^{(s-1)})\pi(\boldsymbol{\theta}^{(s-1)})} \frac{\exp\left[-\frac{1}{2c^2}\sum_{j=1}^{p}\left(\theta_j^{(s-1)} - \theta_j^* + \frac{c^2}{2}\nabla(\boldsymbol{\theta}^*)\right)^2\right]}{\exp\left[-\frac{1}{2c^2}\sum_{j=1}^{p}\left(\theta_j^* - \theta_j^{(s-1)} + \frac{c^2}{2}\nabla(\boldsymbol{\theta}^{(s-1)})\right)^2\right]}.$$

As with the standard Metropolis algorithm, the tuning parameter c should be adjusted to give reasonable acceptance probability. Roberts and Rosenthal [68] argue that 0.574 is the optimal acceptance probability, but they claim that acceptance probabilities between 0.4 and 0.8 work well. In this chapter we have assumed that the candidate standard deviation c is the same for all p parameters and that the candidates are independent across parameters. Convergence can often be improved by adapting the candidate covariance to resemble the posterior covariance. Finally, we note that since MALA is simply a special type of MH sampling it can be used within a larger Gibbs sampling algorithm just like MH sampling steps.

Hamiltonian Monte Carlo (HMC)

MALA improves on random-walk Metropolis sampling by fitting the candidate distribution to the posterior by incorporating the gradient of the log posterior. However, for highly irregular posterior distributions (e.g., a U-shaped or donut-shaped posterior), one step along the gradient may not be sufficient to traverse the posterior. Hybrid Monte Carlo (HMC; also called Hamiltonian Monte Carlo) sampling [60] generalizes MALA to take multiple random steps guided by the gradient. The simple version in Algorithm 4 has two tuning parameters: the step size c and the number of steps L. If $L = 1$ then this algorithm reduces to MALA with c as the candidate standard deviation. Motivation, extensions and tuning of this algorithm are beyond the scope of this text but form the basis for the software STAN [15] which is compared with other MCMC software in Appendix A.5.

Delayed rejection and adaptive Metropolis

Delayed rejection and adaptive Metropolis (DRAM, [39]) is a combination of two ideas: delayed rejection Metropolis [57] and adaptive Metropolis [40]. Adaptive Metropolis allows the covariance of the candidate distribution to evolve across iterations. The intuition is that if the posterior is irregularly shaped, then a different proposal distribution is needed depending on the current state of the chain. Assuming a Gaussian random-walk candidate distribution, $\boldsymbol{\theta}^*|\boldsymbol{\theta}^{(s-1)} \sim \text{Normal}\left(\boldsymbol{\theta}^{(s-1)}, \mathbf{V}^{(s-1)}\right)$, the user sets an initial $p \times p$

Algorithm 4 Hamiltonian MCMC

1: Initialize $\boldsymbol{\theta}^{(0)} = (\theta_1^{(0)}, ..., \theta_p^{(0)})$
2: **for** $s = 1, ..., S$ **do**
3:      sample $\mathbf{z} \sim \text{Normal}(\mathbf{0}, \mathbf{I}_p)$
4:      set $\boldsymbol{\theta}^* = \boldsymbol{\theta}^{(s-1)}$
5:      set $\mathbf{z}^* = \mathbf{z} + c\nabla(\boldsymbol{\theta}^*)/2$
6:      **for** $l = 1, ..., L$ **do**
7:           set $\boldsymbol{\theta}^* = \boldsymbol{\theta}^* + c\mathbf{z}^*$
8:           set $\mathbf{z}^* = \mathbf{z}^* + c\nabla(\boldsymbol{\theta}^*)$
9:      **end for**
10:     set $\mathbf{z}^* = \mathbf{z}^* - c\nabla(\boldsymbol{\theta}^*)/2$
11:     set $R = \dfrac{f(\mathbf{Y}|\boldsymbol{\theta}^*)\pi(\boldsymbol{\theta}^*)}{f(\mathbf{Y}|\boldsymbol{\theta}^{(s-1)})\pi(\boldsymbol{\theta}^{(s-1)})} \cdot \dfrac{\exp(-\mathbf{z}^{*T}\mathbf{z}^*/2)}{\exp(-\mathbf{z}^T\mathbf{z}/2)}$
12:     sample $U \sim \text{Uniform}(0, 1)$
13:     **if** $U < R$ **then**
14:          $\boldsymbol{\theta}^{(s)} = \boldsymbol{\theta}^*$
15:     **else**
16:          $\boldsymbol{\theta}^{(s)} = \boldsymbol{\theta}^{(s-1)}$
17:     **end if**
18: **end for**

---

covariance matrix $\mathbf{V}^{(0)}$ that is then adapted as

$$\mathbf{V}^{(s)} = c\left(\hat{\mathbf{V}}^{(s)} + \delta\mathbf{I}\right)$$

where $\hat{\mathbf{V}}^{(s)}$ is the sample covariance of the previous samples $\boldsymbol{\theta}^{(1)}, ..., \boldsymbol{\theta}^{(s-1)}$, $\delta > 0$ is a small constant to avoid singularities and $c = 2.4^2/p$ [31].

Delayed rejection Metropolis replaces the standard single proposal in Metropolis–Hastings sampling with multiple proposal considered sequentially. The first stage is a usual Metropolis–Hasting step with candidate $\boldsymbol{\theta}^*|\boldsymbol{\theta}^{(s-1)} \sim q\left(\boldsymbol{\theta}^*|\boldsymbol{\theta}^{(s-1)}\right)$ and acceptance probability

$$R\left(\boldsymbol{\theta}^*, \boldsymbol{\theta}^{(s-1)}\right) = \min\left\{1, \frac{p\left(\boldsymbol{\theta}^*|\mathbf{Y}\right)q\left(\boldsymbol{\theta}^{(s-1)}|\boldsymbol{\theta}^*\right)}{p\left(\boldsymbol{\theta}^{(s-1)}|\mathbf{Y}\right)q\left(\boldsymbol{\theta}^*|\boldsymbol{\theta}^{(s-1)}\right)}\right\}.$$

If the first candidate is rejected, a second candidate is proposed as $\boldsymbol{\theta}'|\boldsymbol{\theta}^*, \boldsymbol{\theta}^{(s-1)} \sim Q\left(\boldsymbol{\theta}'|\boldsymbol{\theta}^*, \boldsymbol{\theta}^{(s-1)}\right)$ and accepted with probability

$$\min\left\{1, \frac{p\left(\boldsymbol{\theta}'|\mathbf{Y}\right)q\left(\boldsymbol{\theta}'|\boldsymbol{\theta}^*\right)Q\left(\boldsymbol{\theta}'|\boldsymbol{\theta}^*, \boldsymbol{\theta}^{(s-1)}\right)\left[1 - R\left(\boldsymbol{\theta}', \boldsymbol{\theta}^*\right)\right]}{p\left(\boldsymbol{\theta}^{(s-1)}|\mathbf{Y}\right)q\left(\boldsymbol{\theta}^{(s-1)}|\boldsymbol{\theta}^*\right)Q\left(\boldsymbol{\theta}^{(s-1)}|\boldsymbol{\theta}^*, \boldsymbol{\theta}'\right)\left[1 - R\left(\boldsymbol{\theta}^{(s-1)}, \boldsymbol{\theta}^*\right)\right]}\right\}.$$

The notation becomes cumbersome but this is can be iterated beyond two candidates. DRAM combines these two ideas by using adaptive Metropolis to tune the Gaussian candidate distributions used for q and Q.

Slice sampling

Slice sampling [59] is a clever way to apply Gibbs sampling when the full conditional distributions do not belong to known parametric families of distributions. Slice sampling introduces an auxiliary variable (i.e., a variable that is not an actual parameter) $U > 0$ and draws samples from the joint distribution

$$p^*(\boldsymbol{\theta}, U) = I[0 < U < p(\boldsymbol{\theta}|\mathbf{Y})].$$

By construction, under p^* the marginal distribution of $\boldsymbol{\theta}$ is

$$\int I[0 < U < p(\boldsymbol{\theta}|\mathbf{Y})]dU = p(\boldsymbol{\theta}|\mathbf{Y}),$$

and therefore if samples of $(\boldsymbol{\theta}, U)$ are drawn from p^*, then the samples of $\boldsymbol{\theta}$ follow the posterior distribution. Also, Gibbs sampling can be used to draw samples from p^* since the full conditional distributions are both uniform

1. $U|\boldsymbol{\theta}, \mathbf{Y} \sim$ Uniform on $[0, p(\boldsymbol{\theta}|\mathbf{Y})]$
2. $\boldsymbol{\theta}|U, \mathbf{Y} \sim$ Uniform on $P(U) = \{\boldsymbol{\theta}; p(\boldsymbol{\theta}|\mathbf{Y}) > U\}$

Therefore, slice sampling works by drawing from the joint distribution of $(\boldsymbol{\theta}, U)$, discarding the samples of U and retaining the samples from $\boldsymbol{\theta}$. The most challenging step is to make a draw from $P(U)$ (see the figure below). For some posteriors $P(U)$ has a simple form and samples can be drawn directly. In other cases, $\boldsymbol{\theta}$ can be drawn from a uniform distribution with a domain that includes $P(U)$ until a sample falls in $P(U)$.

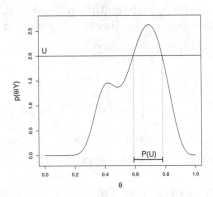

Illustration of slice sampling. The curve is the posterior density $p(\theta|\mathbf{Y})$, the horizontal line represents the auxiliary variable U (i.e., the "slice"), and the bold interval is $P(U) = \{\theta; p(\theta|\mathbf{Y}) > U\}$.

A.5: Software comparison

There are now many software packages to implement the MCMC algorithms discussed in Chapter 3. Here we compare four packages: JAGS, OpenBUGS, STAN and NIMBLE. The packages are all fairly similar to use and so rather than sticking with one package for all analyses, we recommend becoming familiar with multiple packages as different packages will be more effective for some applications than others. As these examples show, JAGS and OpenBUGS are slightly easier to code and sufficiently fast for the low to medium complexity analyses considered in this book, but for more complex models or larger datasets it is useful to be familiar with other packages such as STAN.

Example 1: Simple linear regression

Listings 7.10–7.13 give R code to fit a simple linear regression model using the four packages under consideration. The code is nearly identical for JAGS, OpenBUGS and NIMBLE and slightly more complex for STAN. For this simple example, JAGS has the fastest computation time (see the table below) followed by OpenBUGS, then STAN and then NIMBLE, but this is mostly due to overhead time setting up the more intricate updating schemes in NIMBLE and STAN. The effective sample size for the slope is the highest for STAN and similar for the other three software packages.

MCMC software for simple linear regression. Effective samples size (ESS) and run time (seconds) for two chains each with 30,000 MCMC iterations and a burn-in of 10,000.

Parameter	ESS for β_1	ESS for β_2	Run time
JAGS	4038	4098	0.09
OpenBUGS	1600	1600	3.75
STAN	10391	10360	32.0
NIMBLE	3984	4006	48.0

Example 2: Random slopes model

For a more challenging example, Listings 7.14–7.17 give R code to run a random slopes regression model using the jaw data introduced in Section 4.4 and plotted in Figure 4.6. The table below gives the effective sample size for the means of the random effects distributions (the variances have higher effective sample sizes for all models). As in the simple regression example, JAGS is the fastest package and STAN has the highest effective sample size. Weighing these two factors, in this case STAN is the better option, but all packages are

Listing 7.10
JAGS code for simple linear regression for the paleo data.

```
1    mass <- c(29.9, 1761, 1807, 2984, 3230, 5040, 5654)
2    age  <- c(2, 15, 14, 16, 18, 22, 28)
3    n    <- length(age)
4
5    # Fit in JAGS
6    #install.packages("rjags")
7    library(rjags)
8
9    model_string <- textConnection("model{
10     for(i in 1:n){
11       mass[i] ~ dnorm(beta1 + beta2*age[i],tau)
12     }
13     tau   ~ dgamma(0.01, 0.01)
14     beta1 ~ dnorm(0, 0.0000001)
15     beta2 ~ dnorm(0, 0.0000001)
16   }")
17
18   data  <- list(mass=mass,age=age,n=n)
19   inits <- list(beta1=rnorm(1),beta2=rnorm(1),tau=10)
20   model <- jags.model(model_string, data = data,
            inits=inits,n.chains=2)
21
22   update(model, 10000)
23   samples <- coda.samples(model, n.iter=20000,
24              variable.names=c("beta1","beta2"))
25   summary(samples)
```

Listing 7.11
OpenBUGS code for simple linear regression for the paleo data.

```
1   mass <- c(29.9, 1761, 1807, 2984, 3230, 5040, 5654)
2   age  <- c(2, 15, 14, 16, 18, 22, 28)
3   n    <- length(age)
4
5   #install.packages("R2OpenBUGS")
6   library(R2OpenBUGS)
7
8   model_string <- function() {
9     for(i in 1:n){
10      mass[i] ~ dnorm(mn[i],tau)
11      mn[i] <- beta1 + beta2*age[i]
12    }
13    tau  ~ dgamma(0.01, 0.01)
14    beta1 ~ dnorm(0, 0.0000001)
15    beta2 ~ dnorm(0, 0.0000001)
16  }
17
18  data <- list(mass=mass,age=age,n=n)
19  inits <- list(beta1=rnorm(1),beta2=rnorm(1),tau=10)
20  fit   <- bugs(model.file=model_string,
21                data=data,inits=inits,
22                parameters.to.save=c("beta1","beta2"),
23                n.iter=30000,n.burnin=10000,n.chains=2)
24  fit
```

Listing 7.12
STAN code for simple linear regression for the paleo data.

```
1   mass <- c(29.9, 1761, 1807, 2984, 3230, 5040, 5654)
2   age  <- c(2, 15, 14, 16, 18, 22, 28)
3   n    <- length(age)
4
5   #install.packages("rstan")
6   library(rstan)
7
8   stan_model <- "
9
10    data {
11     int<lower=0> n;
12     vector [n] mass;
13     vector [n] age;
14    }
15
16    parameters {
17      real beta1;
18      real beta2;
19      real<lower=0> sigma;
20    }
21
22    model {
23      vector [n] mu;
24      beta1 ~ normal(0,1000000);
25      beta2 ~ normal(0,1000000);
26      sigma ~ cauchy(0.0,1000);
27      mu   = beta1 + beta2*age;
28      mass ~ normal(mu,sigma);
29    }
30   "
31
32   data <- list(n=n,age=age,mass=mass)
33   fit_stan <- stan(model_code = stan_model,
34                  data = data, chains=2, warmup = 10000, iter =
                        30000)
35   fit_stan
```

Listing 7.13
NIMBLE code for simple linear regression for the paleo data.

```
1   mass <- c(29.9, 1761, 1807, 2984, 3230, 5040, 5654)
2   age  <- c(2, 15, 14, 16, 18, 22, 28)
3   n    <- length(age)
4
5   #install.packages("nimble")
6   library(nimble)
7
8   model_string <- nimbleCode({
9     for(i in 1:n){
10      mass[i] ~ dnorm(mn[i],tau)
11      mn[i] <- beta1 + beta2*age[i]
12    }
13    tau   ~ dgamma(0.01, 0.01)
14    beta1 ~ dnorm(0, 0.0000001)
15    beta2 ~ dnorm(0, 0.0000001)
16  })
17
18  consts  <- list(n=n,age=age)
19  data    <- list(mass=mass)
20  inits   <- function(){list(beta1=rnorm(1),beta2=rnorm(1),tau=10)}
21  samples <- nimbleMCMC(model_string, data = data, inits = inits,
22                        constants=consts,
23                        monitors = c("beta1", "beta2"),
24                        samplesAsCodaMCMC=TRUE,WAIC=FALSE,
25                        niter = 30000, nburnin = 10000, nchains = 2)
26  plot(samples)
27  effectiveSize(samples)
```

Listing 7.14
JAGS code for the random slopes model for the jaw data.

```
1   model_string <- textConnection("model{
2     # Likelihood
3     for(i in 1:n){for(j in 1:m){
4       Y[i,j] ~ dnorm(alpha1[i]+alpha2[i]*age[j],tau3)
5     }}
6
7     # Random effects
8     for(i in 1:n){
9       alpha1[i] ~ dnorm(mu1,tau1)
10      alpha2[i] ~ dnorm(mu2,tau2)
11    }
12
13    # Priors
14    mu1 ~ dnorm(0,0.0001)
15    mu2 ~ dnorm(0,0.0001)
16    tau1 ~ dgamma(0.1,0.1)
17    tau2 ~ dgamma(0.1,0.1)
18    tau3 ~ dgamma(0.1,0.1)
19  }")
20
21  data   <- list(Y=Y,age=age,n=n,m=m)
22  params <- c("mu1","mu2","tau1","tau2","tau3")
23  model  <- jags.model(model_string,data = data,
              n.chains=2,quiet=TRUE)
24  update(model, 10000, progress.bar="none")
25  samples <- coda.samples(model, variable.names=params,
26                          n.iter=90000, progress.bar="none")
27  summary(samples)
```

sufficient, especially if thinning were implemented for JAGS, OpenBUGS and NIMBLE.

MCMC software for the random slopes model. Effective samples size (ESS) and run time (seconds) for two chains each with 100,000 MCMC iterations and a burn-in of 10,000.

Parameter	ESS for μ_1	ESS for μ_2	Run time
JAGS	293	335	1.83
OpenBUGS	1300	1200	10.2
STAN	180000	180000	424
NIMBLE	283	311	26.5

Listing 7.15

OpenBUGS code for the random slopes model for the jaw data.

```
1   model_string <- function(){
2     # Likelihood
3     for(i in 1:n){for(j in 1:m){
4       Y[i,j]   ~ dnorm(mn[i,j],tau3)
5       mn[i,j] <- alpha1[i]+alpha2[i]*age[j]
6     }}
7
8     # Random effects
9     for(i in 1:n){
10      alpha1[i] ~ dnorm(mu1,tau1)
11      alpha2[i] ~ dnorm(mu2,tau2)
12    }
13
14    # Priors
15    mu1 ~ dnorm(0,0.0001)
16    mu2 ~ dnorm(0,0.0001)
17    tau1 ~ dgamma(0.1,0.1)
18    tau2 ~ dgamma(0.1,0.1)
19    tau3 ~ dgamma(0.1,0.1)
20  }
21
22  data    <- list(Y=Y,age=age,n=n,m=m)
23  params <- c("mu1","mu2","tau1","tau2","tau3")
24  inits <- function(){list(mu1=0,mu2=0,tau1=.1,tau2=.2,tau3=.2)}
25  fit   <- bugs(model.file=model_string,
26                data=data,inits=inits,
27                parameters.to.save=params,DIC=FALSE,
28                n.iter=90000,n.chains=2,n.burnin=10000)
29  fit$summary
```

Listing 7.16
STAN code for the random slopes model for the jaw data.

```
1    stan_model <- "
2
3      data {
4        int<lower=0> n;
5        int<lower=0> m;
6        vector [m] age;
7        matrix [n,m] Y;
8      }
9
10     parameters {
11       vector [n] alpha1;
12       vector [n] alpha2;
13       real mu1;
14       real mu2;
15       real<lower=0> sigma3;
16       real<lower=0> sigma2;
17       real<lower=0> sigma1;
18     }
19
20     model {
21       real mu;
22       alpha1 ~ normal(0,sigma1);
23       alpha2 ~ normal(0,sigma2);
24       sigma1 ~ cauchy(0.0,1000);
25       sigma2 ~ cauchy(0.0,1000);
26       sigma3 ~ cauchy(0.0,1000);
27       mu1    ~ normal(0,1000);
28       mu2    ~ normal(0,1000);
29
30       for(i in 1:n){for(j in 1:m){
31         mu     = alpha1[i] + alpha2[i]*age[j];
32         Y[i,j] ~ normal(mu,sigma3);
33       }}
34     }
35   "
36
37   data    <- list(Y=Y,age=age,n=n,m=m)
38   fit_stan <- stan(model_code = stan_model,
39                    data = data,
40                    chains=2, warmup = 10000, iter = 100000)
41   summary(fit_stan)$summary
```

Listing 7.17
NIMBLE code for the random slopes model for the jaw data.

```
1    library(nimble)
2
3    model_string <- nimbleCode({
4      # Likelihood
5      for(i in 1:n){for(j in 1:m){
6        Y[i,j]   ~ dnorm(mn[i,j],tau3)
7        mn[i,j] <- alpha1[i]+alpha2[i]*age[j]
8      }}
9
10     # Random effects
11     for(i in 1:n){
12       alpha1[i] ~ dnorm(mu1,tau1)
13       alpha2[i] ~ dnorm(mu2,tau2)
14     }
15
16     # Priors
17      mu1 ~ dnorm(0,0.0001)
18      mu2 ~ dnorm(0,0.0001)
19     tau1 ~ dgamma(0.1,0.1)
20     tau2 ~ dgamma(0.1,0.1)
21     tau3 ~ dgamma(0.1,0.1)
22   })
23
24   params  <- c("mu1","mu2","tau1","tau2","tau3")
25   consts  <- list(n=n,m=m,age=age)
26   data    <- list(Y=Y)
27   inits   <- function(){
28     list(mu1=rnorm(1),mu2=rnorm(1),tau1=10,tau2=10,tau3=10)
29   }
30   samples <- nimbleMCMC(model_string, data = data, inits = inits,
31                         constants=consts,
32                         monitors = params,
33                         samplesAsCodaMCMC=TRUE,WAIC=FALSE,
34                         niter = 100000, nburnin = 10000, nchains = 2)
35   plot(samples)
```

Bibliography

[1] Helen Abbey. An examination of the Reed-Frost theory of epidemics. *Human Biology*, 24(3):201, 1952.

[2] Hirotogu Akaike. Information theory and an extension of the maximum likelihood principle. In *Selected Papers of Hirotugu Akaike*, pages 199–213. Springer, 1998.

[3] James H Albert and Siddhartha Chib. Bayesian analysis of binary and polychotomous response data. *Journal of the American Statistical Association*, 88(422):669–679, 1993.

[4] Sudipto Banerjee, Bradley P Carlin, and Alan E Gelfand. *Hierarchical Modeling and Analysis for Spatial Data*. CRC Press, 2014.

[5] Albert Barberán, Robert R Dunn, Brian J Reich, Krishna Pacifici, Eric B Laber, Holly L Menninger, James M Morton, Jessica B Henley, Jonathan W Leff, Shelly L Miller, and Noah Fierer. The ecology of microscopic life in household dust. In *Proceedings of the Royal Society B*, volume 282, page 20151139. The Royal Society, 2015.

[6] Maria Maddalena Barbieri and James O Berger. Optimal predictive model selection. *The Annals of Statistics*, 32(3):870–897, 2004.

[7] Daryl J Bem. Feeling the future: Experimental evidence for anomalous retroactive influences on cognition and affect. *Journal of Personality and Social Psychology*, 100(3):407, 2011.

[8] James Berger. The case for objective Bayesian analysis. *Bayesian Analysis*, 1(3):385–402, 2006.

[9] James O Berger, Luis R Pericchi, JK Ghosh, Tapas Samanta, Fulvio De Santis, JO Berger, and LR Pericchi. Objective Bayesian methods for model selection: Introduction and comparison. *Institute of Mathematical Statistics Lecture Notes - Monograph Series*, pages 135–207, 2001.

[10] Jose M Bernardo. Reference posterior distributions for Bayesian inference. *Journal of the Royal Statistical Society: Series B (Methodological)*, pages 113–147, 1979.

[11] Anirban Bhattacharya, Debdeep Pati, Natesh S Pillai, and David B Dunson. Dirichlet–Laplace priors for optimal shrinkage. *Journal of the American Statistical Association*, 110(512):1479–1490, 2015.

[12] Howard D Bondell and Brian J Reich. Consistent high-dimensional Bayesian variable selection via penalized credible regions. *Journal of the American Statistical Association*, 107(500):1610–1624, 2012.

[13] Dennis D Boos and Leonard A Stefanski. *Essential Statistical Inference: Theory and Methods*, volume 120. Springer Science & Business Media, 2013.

[14] Carlos A Botero, Beth Gardner, Kathryn R Kirby, Joseph Bulbulia, Michael C Gavin, and Russell D Gray. The ecology of religious beliefs. *Proceedings of the National Academy of Sciences*, 111(47):16784–16789, 2014.

[15] Bob Carpenter, Andrew Gelman, Matt Hoffman, Daniel Lee, Ben Goodrich, Michael Betancourt, Michael A Brubaker, Jiqiang Guo, Peter Li, and Allen Riddell. STAN: A probabilistic programming language. *Journal of Statistical Software*, 20(2):1–37, 2016.

[16] Carlos M Carvalho, Nicholas G Polson, and James G Scott. The horseshoe estimator for sparse signals. *Biometrika*, 97(2):465–480, 2010.

[17] Fang Chen. Bayesian modeling using the MCMC procedure. In *Proceedings of the SAS Global Forum 2008 Conference, Cary NC: SAS Institute Inc*, 2009.

[18] Hugh A Chipman, Edward I George, and Robert E McCulloch. BART: Bayesian additive regression trees. *The Annals of Applied Statistics*, 4(1):266–298, 2010.

[19] Ciprian M Crainiceanu, David Ruppert, and Matthew P Wand. Bayesian analysis for penalized spline regression using WinBUGS. *Journal of Statistical Software*, 14(1):1–24, 2005.

[20] A Philip Dawid. Present position and potential developments: Some personal views: Statistical theory: The prequential approach. *Journal of the Royal Statistical Society: Series A (General)*, pages 278–292, 1984.

[21] Gustavo de los Campos and Paulino Perez Rodriguez. *BLR: Bayesian Linear Regression*, 2014. R package version 1.4.

[22] Perry de Valpine, Daniel Turek, Christopher J Paciorek, Clifford Anderson-Bergman, Duncan Temple Lang, and Rastislav Bodik. Programming with models: Writing statistical algorithms for general model structures with NIMBLE. *Journal of Computational and Graphical Statistics*, 26(2):403–413, 2017.

[23] Peter Diggle, Rana Moyeed, Barry Rowlingson, and Madeleine Thomson. Childhood malaria in the Gambia: A case-study in model-based geostatistics. *Journal of the Royal Statistical Society: Series C (Applied Statistics)*, 51(4):493–506, 2002.

[24] Stewart M Edie, Peter D Smits, and David Jablonski. Probabilistic models of species discovery and biodiversity comparisons. *Proceedings of the National Academy of Sciences*, 114(14):3666–3671, 2017.

[25] Gregory M Erickson, Peter J Makovicky, Philip J Currie, Mark A Norell, Scott A Yerby, and Christopher A Brochu. Gigantism and comparative life-history parameters of tyrannosaurid dinosaurs. *Nature*, 430(7001):772–775, 2004.

[26] Kevin R. Forward, David Haldane, Duncan Webster, Carolyn Mills, Cherly Brine, and Diane Aylward. A comparison between the Strep A Rapid Test Device and conventional culture for the diagnosis of streptococcal pharyngitis. *Canadian Journal of Infectious Diseases and Medical Microbiology*, 17:221–223, 2004.

[27] Seymour Geisser. Discussion on sampling and Bayes inference in scientific modeling and robustness (by GEP Box). *Journal of the Royal Statistical Society: Series A (General)*, 143:416–417, 1980.

[28] Alan E Gelfand, Peter Diggle, Peter Guttorp, and Montserrat Fuentes. *Handbook of Spatial Statistics*. CRC press, 2010.

[29] Andrew Gelman et al. Prior distributions for variance parameters in hierarchical models (comment on article by Browne and Draper). *Bayesian Analysis*, 1(3):515–534, 2006.

[30] Andrew Gelman, Jessica Hwang, and Aki Vehtari. Understanding predictive information criteria for Bayesian models. *Statistics and Computing*, 24(6):997–1016, 2014.

[31] Andrew Gelman, Gareth O Roberts, and Walter R Gilks. Efficient Metropolis jumping rules. *Bayesian Statistics*, 5(599-608):42, 1996.

[32] Andrew Gelman and Donald B Rubin. Inference from iterative simulation using multiple sequences. *Statistical Science*, pages 457–472, 1992.

[33] Andrew Gelman and Cosma Rohilla Shalizi. Philosophy and the practice of Bayesian statistics. *British Journal of Mathematical and Statistical Psychology*, 66(1):8–38, 2013.

[34] Stuart Geman and Donald Geman. Stochastic relaxation, Gibbs distributions, and the Bayesian restoration of images. In *Readings in Computer Vision*, pages 564–584. Elsevier, 1987.

[35] Edward I George and Robert E McCulloch. Variable selection via Gibbs sampling. *Journal of the American Statistical Association*, 88(423):881–889, 1993.

[36] John Geweke. *Evaluating the accuracy of sampling-based approaches to the calculation of posterior moments*, volume 196. Federal Reserve Bank of Minneapolis, Research Department, Minneapolis, MN, USA, 1991.

[37] Subhashis Ghosal and Aad van der Vaart. *Fundamentals of Nonparametric Bayesian Inference*, volume 44. Cambridge University Press, 2017.

[38] Tilmann Gneiting and Adrian E Raftery. Strictly proper scoring rules, prediction, and estimation. *Journal of the American Statistical Association*, 102(477):359–378, 2007.

[39] Heikki Haario, Marko Laine, Antonietta Mira, and Eero Saksman. DRAM: Efficient adaptive MCMC. *Statistics and Computing*, 16(4):339–354, 2006.

[40] Heikki Haario, Eero Saksman, and Johanna Tamminen. An adaptive Metropolis algorithm. *Bernoulli*, 7(2):223–242, 2001.

[41] Wilfred K Hastings. Monte Carlo sampling methods using Markov chains and their applications. *Biometrika*, 57(1):97–109, 1970.

[42] James S Hodges. *Richly Parameterized Linear Models: Additive, Time Series, and Spatial Models using Random Effects*. Chapman & Hall/CRC, 2016.

[43] James S Hodges and Brian J Reich. Adding spatially-correlated errors can mess up the fixed effect you love. *The American Statistician*, 64(4):325–334, 2010.

[44] Arthur E Hoerl and Robert W Kennard. Ridge regression: Biased estimation for nonorthogonal problems. *Technometrics*, 12(1):55–67, 1970.

[45] Jennifer A Hoeting, David Madigan, Adrian E Raftery, and Chris T Volinsky. Bayesian model averaging: A tutorial. *Statistical Science*, pages 382–401, 1999.

[46] Michael I Jordan, Zoubin Ghahramani, Tommi S Jaakkola, and Lawrence K Saul. An introduction to variational methods for graphical models. *Machine Learning*, 37(2):183–233, 1999.

[47] Bindu Kalesan, Matthew E Mobily, Olivia Keiser, Jeffrey A Fagan, and Sandro Galea. Firearm legislation and firearm mortality in the USA: A cross-sectional, state-level study. *The Lancet*, 387(10030):1847–1855, 2016.

[48] Robert E Kass and Adrian E Raftery. Bayes factors. *Journal of the American Statistical Association*, 90(430):773–795, 1995.

[49] Robert E Kass and Larry Wasserman. A reference Bayesian test for nested hypotheses and its relationship to the Schwarz criterion. *Journal of the American Statistical Association*, 90(431):928–934, 1995.

[50] Hong Lan, Meng Chen, Jessica B Flowers, Brian S Yandell, Donnie S Stapleton, Christine M Mata, Eric Ton-Keen Mui, Matthew T Flowers, Kathryn L Schueler, Kenneth F Manly, et al. Combined expression trait correlations and expression quantitative trait locus mapping. *PLoS Genetics*, 2(1):e6, 2006.

[51] Dennis V Lindley. A statistical paradox. *Biometrika*, 44(1/2):187–192, 1957.

[52] Roderick Little. Calibrated Bayes, for statistics in general, and missing data in particular. *Statistical Science*, 26(2):162–174, 2011.

[53] Jean-Michel Marin, Pierre Pudlo, Christian P Robert, and Robin J Ryder. Approximate Bayesian computational methods. *Statistics and Computing*, 22(6):1167–1180, 2012.

[54] Peter McCullagh and John Nelder. *Generalized Linear Models, Second Edition*. Boca Raton: Chapman & Hall/CRC, 1989.

[55] Nicholas Metropolis, Arianna W Rosenbluth, Marshall N Rosenbluth, Augusta H Teller, and Edward Teller. Equation of state calculations by fast computing machines. *The Journal of Chemical Physics*, 21(6):1087–1092, 1953.

[56] Greg Miller. ESP paper rekindles discussion about statistics. *Science*, 331(6015):272–273, 2011.

[57] Antonietta Mira. On Metropolis-Hastings algorithms with delayed rejection. *Metron*, 59(3-4):231–241, 2001.

[58] Frederick Mosteller and David L Wallace. Inference in an authorship problem: A comparative study of discrimination methods applied to the authorship of the disputed Federalist Papers. *Journal of the American Statistical Association*, 58(302):275–309, 1963.

[59] Radford M Neal. Slice sampling. *Annals of Statistics*, pages 705–741, 2003.

[60] Radford M Neal. MCMC using Hamiltonian dynamics. *Handbook of Markov Chain Monte Carlo*, 2(11):2, 2011.

[61] Radford M Neal. *Bayesian Learning for Neural Networks*, volume 118. Springer Science & Business Media, 2012.

[62] Jorge Nocedal and Stephen J Wright. *Sequential Quadratic Programming.* Springer, 2006.

[63] Krishna Pacifici, Brian J Reich, David AW Miller, Beth Gardner, Glenn Stauffer, Susheela Singh, Alexa McKerrow, and Jaime A Collazo. Integrating multiple data sources in species distribution modeling: A framework for data fusion. *Ecology*, 98(3):840–850, 2017.

[64] Anand Patil, David Huard, and Christopher J Fonnesbeck. PyMC: Bayesian stochastic modelling in Python. *Journal of Statistical Software*, 35(4):1, 2010.

[65] LI Pettit. The conditional predictive ordinate for the normal distribution. *Journal of the Royal Statistical Society: Series B (Methodological)*, pages 175–184, 1990.

[66] Martyn Plummer. JAGS Version 4.0. 0 user manual. *See https://sourceforge. net/projects/mcmc-jags/files/Manuals/4. x*, 2015.

[67] Carl Edward Rasmussen. Gaussian processes in machine learning. In *Advanced Lectures on Machine Learning*, pages 63–71. Springer, 2004.

[68] Gareth O Roberts and Jeffrey S Rosenthal. Optimal scaling of discrete approximations to Langevin diffusions. *Journal of the Royal Statistical Society: Series B (Statistical Methodology)*, 60(1):255–268, 1998.

[69] Gareth O Roberts and Adrian FM Smith. Simple conditions for the convergence of the Gibbs sampler and Metropolis-Hastings algorithms. *Stochastic Processes and Their Applications*, 49(2):207–216, 1994.

[70] Veronika Ročková and Edward I George. The spike-and-slab LASSO. *Journal of the American Statistical Association*, 113(521):431–444, 2018.

[71] Donald B Rubin. Bayesianly justifiable and relevant frequency calculations for the applied statistician. *The Annals of Statistics*, pages 1151–1172, 1984.

[72] Donald B Rubin. Multiple imputation after 18+ years. *Journal of the American Statistical Association*, 91(434):473–489, 1996.

[73] Håvard Rue, Sara Martino, and Nicolas Chopin. Approximate Bayesian inference for latent Gaussian models by using integrated nested Laplace approximations. *Journal of the Royal Statistical Society: Series B (Statistical Methodology)*, 71(2):319–392, 2009.

[74] John R Sauer, James E Hines, and Jane E Fallon. *The North American Breeding Bird Survey, Results and Analysis 1966–2005*. Version 6.2.2006. USGS Patuxent Wildlife Research Center, Laurel, Maryland, USA, 2005.

[75] Steven L Scott, Alexander W Blocker, Fernando V Bonassi, Hugh A Chipman, Edward I George, and Robert E McCulloch. Bayes and big data: The consensus Monte Carlo algorithm. *International Journal of Management Science and Engineering Management*, 11(2):78–88, 2016.

[76] Daniel Simpson, Håvard Rue, Andrea Riebler, Thiago G Martins, and Sigrunn H Sørbye. Penalising model component complexity: A principled, practical approach to constructing priors. *Statistical Science*, 32(1):1–28, 2017.

[77] David J Spiegelhalter, Nicola G Best, Bradley P Carlin, and Angelika Van Der Linde. Bayesian measures of model complexity and fit. *Journal of the Royal Statistical Society: Series B (Statistical Methodology)*, 64(4):583–639, 2002.

[78] Mervyn Stone. Necessary and sufficient condition for convergence in probability to invariant posterior distributions. *The Annals of Mathematical Statistics*, pages 1349–1353, 1970.

[79] Sibylle Sturtz, Uwe Ligges, and Andrew Gelman. R2OpenBUGS: A package for running OpenBUGS from R. *URL http://cran. rproject. org/web/packages/R2OpenBUGS/vignettes/R2OpenBUGS. pdf*, 2010.

[80] Brian L Sullivan, Christopher L Wood, Marshall J Iliff, Rick E Bonney, Daniel Fink, and Steve Kelling. eBird: A citizen-based bird observation network in the biological sciences. *Biological Conservation*, 142(10):2282–2292, 2009.

[81] Robert Tibshirani. Regression shrinkage and selection via the LASSO. *Journal of the Royal Statistical Society: Series B (Methodological)*, pages 267–288, 1996.

[82] Luke Tierney. Markov chains for exploring posterior distributions. *Annals of Statistics*, pages 1701–1728, 1994.

[83] Arnold Zellner. On assessing prior distributions and Bayesian regression analysis with g-prior distributions. *Bayesian Inference and Decision Techniques*, 1986.

[84] Yan Zhang, Brian J Reich, and Howard D Bondell. High dimensional linear regression via the R2-D2 shrinkage prior. *arXiv preprint arXiv:1609.00046*, 2016.

Index

Absolute loss, 218
Adaptive Metropolis, 252
Akaike information criterion (AIC),
    177
Alternative hypothesis, 27, 219
Aperiodic, 246
Assumptions, 149
Asymptotics, 220, 223
Autocorrelation, 104
Autoregressive model, 156
Auxiliary variable, 254

Basis expansion, 202
Basis functions, 151
Bayes factors, 166
Bayes rule, 218
Bayes' theorem, 14, 21
Bayesian analysis, 18
Bayesian central limit theorem, 135,
    223
Bayesian decision theory, 218
Bayesian information criteria (BIC),
    177
Bayesian model averaging, 176
Bayesian network, 196
Bayesian risk, 218
Bernoulli distribution, 4, 135
Beta distribution, 9, 140
Bias, 220, 224
Bias-variance tradeoff, 221
Big data, 110
Binary data, 135
Binomial distribution, 4, 139
Blocked Gibbs sampling, 87

Candidate distribution, 89
Check loss, 219

Classification, 220
Clustering, 153
Collinearity, 127
Conditional distribution, 10
Confidence interval, 26
Consistent, 220
Continuous random variable, 2
Convergence, 81, 108, 223
Convergence diagnostics, 103
Correlated data, 144, 155
Correlation, 11
Count data, 137
Covariance, 11
Credible interval, 26
Cross validation, 164, 187
Cumulative distribution function, 6

Data fusion, 200
Data layer, 196
Delayed rejection and adaptive
    Metropolis (DRAM), 252
Deviance information criteria (DIC),
    177, 202, 207
Diagnostics, 187
Directed acyclic graph (DAG), 195
Dirichlet process, 153
Discrete random variable, 2
Double exponential, 128

Effective sample size, 106
Empirical Bayes, 62
Equal-tailed interval, 26
Estimator, 25
Exchangeability, 148
Expected value, 3, 6, 11

Fixed effect, 141
Frequentist, 2, 23

Printed in the United States
by Baker & Taylor Publisher Services